CAMBRIDGE COUNTY GEOGRAPHIES

General Editor: F. H. H. GUILLEMARD, M.A., M.D.

FLINTSHIRE

Cambridge County Geographies

FLINTSHIRE

by

J. M. EDWARDS, M.A. (Oxon.)

Head Master of the Holywell County School

With Maps, Diagrams and Illustrations

Cambridge:
at the University Press
1914

CAMBRIDGE UNIVERSITY PRESS
Cambridge, New York, Melbourne, Madrid, Cape Town,
Singapore, São Paulo, Delhi, Mexico City

Cambridge University Press
The Edinburgh Building, Cambridge CB2 8RU, UK

Published in the United States of America by Cambridge University Press, New York

www.cambridge.org
Information on this title: www.cambridge.org/9781107664067

First published 1914
First paperback edition 2013

A catalogue record for this publication is available from the British Library

ISBN 978-1-107-66406-7 Paperback

PREFACE

THIS little volume contains but a brief outline of the story of Flintshire. It is a long and interesting narrative when fully and rightly told.

The undulating plateau of the County was the scene of successive struggles between the men of Gwynedd and their invaders,—the Roman, the Saxon, the Dane, the Norman, Henry II and Edward I.

In later times the Mostyns, the Glynnes,—two of its most illustrious families,—helped to place the Welshman, Henry Tudor, on the throne of England, to bring peace and prosperity, instead of dynastic war.

In still later times its great mineral wealth was discovered, and its Geology became as interesting as its History.

No complete History of the County has been written, but the historian of the future, when he undertakes the task, will find much help in the works of Mr Henry Taylor and Professor Tout, while the records of the naturalist and traveller Thomas Pennant contain much valuable material. More recent information (to which I am greatly indebted) is to be found in Mr A. Strahan's volume on the Geology of the County, and in the Report of the Royal Commission on its Antiquities.

In collecting material I had to tax the patience of many—I now express my sincerest gratitude to all. In particular I wish to thank Mr J. Philip Jones, J.P., Holywell, for much kindly help, especially on Geology, and to Mr A. A. Dallman, F.C.S., Wallasey, for the interesting chapter on Natural History.

J. M. EDWARDS.

July 1914.

CONTENTS

CONTENTS

ILLUSTRATIONS

MAPS AND DIAGRAMS

The illustrations on pp. 15, 17, 18, 24, 36, 62, 66, 68, 75, 79, 121, 136, 145 and 165, are from photographs by F. & M. Davies, Mold; those on pp. 3, 99, 102, 110, 114, 128 and 161, by Mr W. M. Dodson, Bettws-y-Coed; those on pp. 22, 50, 118 and 120, by Messrs Frith & Co.; those on pp. 7, 27, 132 and 138, by Mr C. G. Caldecott, Wrexham; those on pp. 10, 52, 54, 147 and 160, by Mr and Mrs Williams, Rhyl; those on pp. 26, 88, 93 and 122, by Mr J. Thomas, Cambrian Gallery; those on pp. 85, 86 and 135, by Mr G. Mark Cook, Chester; those on pp. 5 and 83 by Mr A. J. Martyn; those on pp. 42 and 44 by Mr A. A. Dallman, F.C.S., Wallasey; those on pp. 139 and 162 by Mr W. Bell Jones, Hawarden; that on p. 12 by Capt. E. L. Marriott, R.N.R., Connah's Quay.

That on p. 73 is from a photograph supplied by Messrs I. J. Abdela & Mitchell, Ltd.; that on p. 74 is by Mr T. Waterhouse, Holywell; that on p. 109 is by Mr J. H. Morris, West Bromwich; that on p. 123 is by Mr R. Newstead, Chester; that on p. 133 is by Dr R. Geoffrey Williams, Wrexham; that on p. 155 is from a portrait in the possession of the Royal Geographical Society, by kind permission of the Society.

1. County and Shire. Meaning and Origin of the word Flint.

Before the Norman and Edwardian conquests Wales was ruled by its own Princes, and the country was divided into areas known as Cantrevs and Commotes. One of the chief aims of Edward I was to annex Wales to England; and as William the Conqueror had demolished the great Earldoms in England and given smaller and scattered portions of land to his barons, lest they should become too powerful, so Edward I also saw that, before he could effectively govern the country, the three Welsh principalities of Gwynedd, Powys, and Deheubarth must be divided into areas similar to English shires and brought directly under his own administration. Accordingly, in March, 1284, he held a Parliament at Rhuddlan, when the important Statute of Wales was passed. This Statute provided that the Principality of Wales should be divided into six counties:—Flintshire, Carnarvonshire, Anglesey, Merionethshire, Cardiganshire, and Carmarthenshire. The area of the county of Flint, when first made a shire, was much smaller than it is now; it consisted of little more than the modern commotes of Coleshill, Prestatyn, and Rhuddlan—practically the old division of Tegeingl.

The English word *shire* has long been used to denote a large division of the country *shorn off*, or separated by boundaries from the rest of the land. It is a *share*, to use another related form of the Early English word, of a previous wider area. The Welsh word "Sir" is derived from the word "shire." "Swydd" is also used, and it is correct to write either "Swydd Flint" or "Sir Flint" for Flintshire. When the Normans came, they introduced the word "county" (*Comté*), a name given to a district ruled over by a Count (*Comte*). This word thus came to be applied to the shire, and nowadays the words "County" and "Shire" are almost identical in meaning in ordinary conversation and literature.

Some of the Welsh counties took their names from an important town in the district, such as Carnarvon and Pembroke, while others, like Merioneth, are called after some ancient division that forms the nucleus of the shire. Flintshire belongs to the former, for it adopted the name of a fort and a town which played an important part in its history.

Several derivations of the word Flint have been suggested. But before the end of the thirteenth century there is no record of the word, although reference is made in Domesday Book to Croesati and Coleshill, two places on either side of the town. There was a ford across the Dee at low tide, where the ruins of Flint castle now stand, and, as it was important to secure this passage, Edward I built a castle, or fortified the one already in existence, known as *Castrum apud fluentum*, i.e. the Castle by the river. Gradually, no doubt, houses were built under the shadow of the castle, and there are records alluding to the town in the reign of Edward I as *apud le Flynt*, and later the word *Flint* was used. Some authorities, however, have suggested that it is merely the ordinary English word "flint" (silex). In the Calendar of Welsh Rolls (1277) it is called Le Chaylon from a French word meaning rock.

The Welsh name for Flintshire in the thirteenth century was Tegeingl. This is derived from the name

Roman Pig of Lead stamped Deceangi

of the inhabitants of these parts before the advent of the Romans—the Deceangi. The word "Deceangi," which was found stamped on several Roman pigs of lead discovered in the district, gradually changed into Tegeingl. When the Mercians conquered Tegeingl, they called it Englefield, which is interpreted by some to mean "the land of the English." It is far more probable, however, that they borrowed *Eingl*, part of the old Celtic word Tegeingl, and added to it the suffix *field* to form Englefield.

2. General Characteristics.

Although Flintshire is the smallest of all the Welsh counties, it is in proportion to its size the richest in minerals and the most densely populated, with the single exception of Glamorganshire.

It is a maritime county and has good soil ; but it is not the nearness of the sea or the quality of the land that has made Flintshire what it is, but its coal and lead. Ever since the time of the Romans lead has been successfully worked in it, and as the productive coal measures form a belt round the county, it is but natural that the mines should provide one of its chief industries, and we find that over three thousand persons earn their living in them.

The geographical position of Flintshire is favourable to the easy and inexpensive distribution of its mineral wealth. It is situated on the border of England, and it has easy access to Liverpool and the Midland counties. Three of the leading railway companies serve it, and when the Dee Conservancy Board completes its important scheme for the navigation of the Dee, the coast from Saltney to Rhyl will become one of the most busy industrial centres in Britain. Even now it can show many thriving industries.

Flintshire can still be called a well wooded county although the dense forest that grew in the Middle Ages no longer exists. The huge trunks of trees exposed along the coast and in Whixall Moss bear evidence of the existence of a forest that once covered the entire county

Submerged Forest

(*This view is taken at Dove Point on the Cheshire coast*)

with the exception possibly of its hill tops. Edward I took ten days to lead his army from Chester to Flint, as he had to work his way through the forest. Local place-names also testify to the widespread existence of wooded country, such as Coed Llai (Llai's Wood), Coed Talon, Helygain, Ysceifiog, Coed Du, and Coed yr Esgob, Llys-y-Coed, and Llwyni.

The natural scenery of Flint has its own characteristics. It cannot boast of high mountains like those of Arvon, or large lakes and steep precipices like those of Merioneth. The coast has none of the rugged grandeur of the rocky seaboard of Pembrokeshire. But no county possesses more beautiful dingles and glens, watered by brooks, along the sides of which steep and zigzag paths wind through shady trees, commanding varied and picturesque views at every turn. In this type of scenery it resembles Devonshire, and in some respects Somersetshire. On the west lies the Vale of Clwyd, the "Paradise of Wales." On the south-east is Maelor, the detached portion, where "gentle Deva flows" under overhanging trees, to form one of the most beautiful parts of Britain.

Forming, as it does, a plateau between England and Wales, the view from some of the hills over land and sea is extensive. On a clear day we can see Cheshire, Lancashire, Cumberland, Westmorland, York (Ingleborough Hill), Derbyshire, and the Isle of Man, and under favourable conditions even Ireland and the headland of Kirkcudbright in Scotland are visible, while southwards over the Denbighshire hills the Merionethshire mountains and the Snowdonian range rise majestically.

The hills that run from Prestatyn through Ysceifiog and Rhosesmor are composed of limestone, and easily absorb the surface water. This is the reason for one of the chief characteristics of the county, namely the number of its springs. The county has no lakes of the size of Bala Lake or Lake Vyrnwy in the neighbouring counties

The Dee in Maelor

of Merioneth and Montgomery. Its most extensive inland sheets of water are the meres of Maelor and Llyn Helyg. As one would expect from its size and the nature of its geological formation, there are no large rivers in Flintshire with the exception of the Dee.

Flintshire figures prominently in Welsh history; almost every important incident in the annals of the northern parts of Wales is closely connected with it from the time when the Roman sentinels watched its hills from the walls of Chester until Henry Tudor summoned its chiefs to join him in his invasion of England. As Shrewsbury and the Severn afforded an entrance into mid-Wales, so Chester and the estuary of the Dee served as a pathway into the northern parts of the Principality. The Romans after building an altar to some unknown goddess, whom they called Deva, at the mouth of the Dee, constructed a road through the forests of Flint westwards to Arvon to conquer and to administer. Guided by like geographical reasons the modern engineer made his railway along the same route two thousand years later to further trade and commerce.

And thus, as Flintshire was one of the gateways of Wales, it became of necessity one of its great battle-grounds where the strength of Gwynedd gathered to oppose the army of invader after invader. It is no wonder, then, that the military remains, especially of castles, are so numerous, and that we so often find an echo of war and sorrow and slaughter in the local place-names. The "Hill of Arrows," the "Hill of Slaughter," the "Blood-drenched Moor," the "Hollow of Woe," the "Hill of the Chariots," the "Hillock of Contention," the "Hill of Execution," the "Hill of Lamentation," are names that vividly suggest some of the story of Wales.

3. Size. Shape. Boundaries.

Flintshire is a maritime county of small size on the north coast of Wales. It is bounded on the north by the Irish Sea, on the north-east by the estuary of the Dee, on the east for a small distance by Cheshire, and on the south and south-west by Denbighshire.

It measures about 30 miles in length and 15 miles in breadth, and is about 115 miles in circumference. The mainland has been calculated to contain 164,774 acres or 257 square miles. The neighbouring counties, Denbighshire and Merionethshire, are more than 2½ times its size, and it would require three Flintshires to equal Montgomeryshire, while Carmarthenshire is more than 3½ times as big.

Lying with its long axis nearly north-west and south-east, the shape of Flintshire may perhaps be best described as a parallelogram expanding into T-shaped form at its south-eastern end, and with the southern border somewhat irregular in outline. The coast is very even and flat; indeed, with the exception of a few sand-hills that rise 40 to 50 feet above sea level between Point of Ayr and Rhyl, the land is almost level with the sea.

The boundary line of Flintshire, like that of all the Welsh counties, is exceedingly irregular. If we look at a map of North America we find that the line of division between the various States is straight. America is a comparatively new country and no difficulty was experienced in plotting out the land, but the cantrevs and commotes of Wales were formed centuries ago, and every curve in the boundary line speaks of some geographical or other difficulty. We will now proceed to trace it in detail, starting from the extreme north-western point near Rhyl.

The boundary commences with the estuary of the Clwyd, east of Foryd, "the ford of the sea," and proceeds along the river for about a mile, where it deviates south

The River Clwyd

and west in a zigzag line till it reaches the four cross roads on Rhuddlan marsh. Here it strikes into the Rhuddlan-Abergele road, which it follows for a quarter of a mile west, when it turns south and cuts across the Bodelwyddan road, proceeding under St George into Kinmel Park. It then enters the road to St Asaph *via* Bryncelyn, and follows it for nearly 2½ miles, when it makes a short loop to the south, crosses the road again, and then turns once more southward *via* Bryn Elwy, crossing the Elwy to reach the river Clwyd just north of Llannerch Park. It follows this river as far as Pont Ruffydd, a distance of some three or four miles, and then turns up the Wheeler towards Bodfari, and so to beyond Pentre-uchaf, where it makes a sudden bend north and as suddenly again returns to Afonwen, just including Caerwys. Thence its course lies along the Clwydian range over Moel-y-Parc and Pen-y-Cloddiau, Moel Arthur and Moel Llys-y-Coed, Penmachno, and Moel Dywyll, to the north-east summit of Moel Fammau, where it diverges east from the Clwydian range and crosses the Leet below Loggerheads. Leaving Cat Hole on the Mold and Ruthin road the boundary proceeds southwards in a meandering line over Moel Findeg, Nerquis Mountain, and Mynydd Du, to Rhyd-y-Ceirw, "the ford of the stags," on the Terrig river. Some distance after passing Rhyd-talog it makes a junction with Nant-y-Ffrith which it follows through romantic scenery into the Cegidog river.

The line now follows the river for Cefn-y-Bedd, enclosing the narrow strip of land formed in the angle of the Cegidog and the Alun. Leaving Caergwrle, Hope, and Caer Estyn a mile or so within it to the west, it follows the Pulford brook and then turns sharp north towards Kinnerton station, comes close to Bretton Hall, and curves round Saltney, where, close to the station, it crosses the Dee. From that point in a more or less north-westerly direction it traverses the reclaimed land of the district of Sealand in the old bed of the river Dee before the diversion of the river in the latter part of the eighteenth century, passing Saughall station and Shotwick Castle and reaching the Dee estuary below Burton Point.

On the north-east the Dee separates it from the Wirral Peninsula. The estuary is two miles across at Connah's Quay, and it widens gradually as we approach

Connah's Quay

the sea. From Point of Ayr to Hoylake it measures about six miles. Some sixty years ago it was possible for people to walk from Flint across the estuary to Parkgate when the tide was out, and considerable use was made of this track. As we glance across we are deceived as to its width and many an unfortunate wanderer has met an untimely death through being caught by the tide on these trackless wastes. Charles Kingsley, in his famous poem,

"The Sands of Dee," has vividly described the return of the treacherous tide, driven by the western wind "wild and dank with foam":

> "The creeping tide came up along the sand,
> And o'er and o'er the sand,
> And round and round the sand,
> As far as eye could see.
> The blinding mist came down and hid the land,
> But never home came she."

Maelor Saesneg, a detached portion of the county, over ten miles distant from its nearest point, lies on the east side of the Dee, and is about ten miles in length by six in breadth. It is almost entirely an agricultural district and is famous for its picturesque scenery, more especially that part of it which lies along the banks of the Dee. It was incorporated by Edward I with the new shire of Flint on its formation in 1284. It is bounded on the north-east by Cheshire; the Dee on the west separates it from Denbighshire, whilst on the south Whixall Moss links it to Shropshire.

In Domesday Book Maelor Saesneg is included in the hundred of Dudestan, when it may have been part of the Mercian diocese of Lichfield. It remained in that of Cheshire until the year 1849. As the other portion of Maelor on the west side of the Dee belonged to the Welsh diocese of St Asaph, this may have been the reason why the former was called Maelor Saesneg—English Maelor, and the latter Maelor Gymreig—Welsh Maelor.

Near Rossett, where the Alun bends almost at right

angles, there is another small piece of land belonging to Flintshire—Marford and Hoseley. These two townships of about 604 acres composed the manor of the same name, Hoseley forming the southern and Marford the northern portion of the manor. They are surrounded by Denbighshire but belong to the parish of Hope in this county. Hoseley was a part of Cheshire in 1087 and belonged to the monastery of St Werburgh. In Domesday Book it appears as "Odeslei"—Oda's Lea. It is probable that these two townships were attached to Flintshire to meet the demands of the powerful Stanley family to have all their territory in the same county. Now they form a connecting link between the mainland of Flintshire and the hundred of Maelor beyond the Dee. Henry VIII had Hope, Mold, Hawarden, and Bromfield manors included in the county of Flint.

4. Surface and General Features.

Referring to the physical map we notice that, except on the south, the county is surrounded by a fringe of land which is nowhere above 300 feet in height; most of it, indeed, being less than 100 feet. With a rise of 50 feet in the ocean bed St Asaph would take the place of Rhyl as a seaside resort. From the lowland the central plain of the county rises very suddenly, owing to geological disturbances in the far past, especially along the coast from Holywell to Diserth, up the vale to Tremeirchion, and south-east to the valley of the Wheeler towards

Afonwen and Nannerch. Pen-y-Ball, near Holywell, although only two miles distant from the sea, is 800 feet high, and so steep is the road to its summit that it is becoming famous as a test for motor-cyclists.

The plateau that extends from Newmarket towards Rhosesmor has an average height of about 500 feet and is composed of limestone. This district is honeycombed

Moel Fammau

with old lead-mines, and the derelict shafts tell tales of fortunes won and lost. The land thence slopes gradually eastward into the plain of Cheshire and the estuary of the Dee.

A valley which forms the only gap in the Clwydian range from Prestatyn to Corwen bisects the county. It runs from Bodfari to Caergwrle and is the basin of the rivers Alun and Wheeler.

The highest mountain is Moel Fammau,—"Mother of Hills," 1819 feet. Several other hills in the Clwydian range exceed 1000 feet, Moel-y-Parc, Pen-y-Cloddiau (1442), Moel Arthur (1494), and Moel Llys-y-Coed (1524), as well as Moel Findeg (1197), Nerquis Mountain (1239) and Hope Mountain (1080) towards the eastern boundary.

During the Glacial Period what is now modern Flintshire extended far into the ocean. Six great glaciers ran from Snowdon in various directions. One of these reached our county, as we know from finding isolated blocks of granite deposited by it. Another huge glacier from the north met this, as the "erratics" or boulders carried from Scafell testify. These boulders are scattered over the county. They abound on the slopes of Moel-y-Parc, Moel Arthur, and Ffrith-y-Garreg Wen, and are of various sizes, one measuring 18 by 10 by 6 feet. The line between the western and the northern boulders may be roughly drawn along the eastern border of Halkyn mountain and thence by Caergwrle to Denbighshire. The county then formed the watershed between North Wales and the Cumberland hills. The glaciers have left their mark on the rocks, especially on Moel Hiraddug, Cwm Mountains, and at Pen-y-Gelli near Lloc, where the rock-surface in places is well polished and scratched. The beautiful valleys and glens of the county were to a very great extent moulded by these "rivers of ice."

The "Drift" in the county travelled from west and south-west, while that of Cheshire came down from the north and north-west. It is interesting therefore to notice that Flintshire lies on the boundary along which the Drift from the west meets the Drift from the north. The glacial deposit of the valley of the Alun and the Halkyn range west and south of Holywell travelled from the west, whilst that along the coast came from the north. It was in the Glacial Period that the Red Sandstone of the Vale of Clwyd travelled up the valley of the Wheeler, and a

Valley of the Alun

remarkable change then took place in the physical geography of the district. Several hillocks and rounded outlines of sand-hills were formed which are still prominent features for several miles along the valley. The hills are called 'eskers' or 'bryniau,' and have often been mistaken for British and Roman earthworks. One of the most famous drift 'eskers' in the county is Bailey Hill at Mold, which is a fine example of sand and gravel esker, though its original form has been artificially modified. On Halkyn Mountain the Drift attains a high altitude and forms many eskers, of which the most conspicuous in the district is Moel-y-Crio (928 feet).

The long hill from Hope to Hawarden is entirely overspread by Drift and Boulder Clay with irregular outcrops of sand and gravel.

Bailey Hill

5. Watershed—Rivers and Lakes.

The watershed of Flintshire needs no special consideration as the natural drainage is comparatively simple, though not always well defined. A ridge of Carboniferous Limestone runs from Prestatyn right along the county through Halkyn and Buckley to Hope. The streams that flow from the numerous springs on the plateau of the watershed run to the Clwyd, or to the estuary of the Dee. Some of them flow to the Mold and Caerwys valley on their way to the Dee or Clwyd. The latter valley forms a remarkable physical feature in the formation of the district, in it the Alun carries the streams south-east into the Dee, or the Wheeler takes them north-west into the Clwyd.

The watershed between the Clwyd and the Alun follows the summit level of the Silurian hills above Caerwys, Moel-y-Parc, and Pen-y-Cloddiau as far as Moel Arthur. From this place it turns in a north-eastward direction to Moel Plas Yw, and follows the road by Bryn-y-Groes to the south side of Penbedw Park. Here it crosses the head of the Nannerch valley, and it is difficult to define exactly where the line of watershed is. At times, indeed, the water runs in both directions, thus at Siamber Wen the water enters a subterranean course (as the Alun does) and reappears at a place called Tardd-y-Dwr "the spring of water," about six hundred yards distant, and flows towards Rhyd-y-Mwyn, "the ford of the ore," where it joins the Alun, but after heavy rain the water that takes the open air course runs into the Wheeler. Thus the water of the stream enters the sea either near Rhyl along the Clwyd, or near Chester along the Dee. The watershed from this point runs north-eastwards to Moel Ffagnallt, through Rhes-y-cae towards the Chert hill of Pentre Halkyn and Pantasaph.

Two of the most beautiful rivers in Wales—the

Dee and the Clwyd—find their way to the sea through Flintshire.

The Dee, "the wizard stream" of Milton and the "sacred Dee" of Tennyson, has been a favourite subject of both English and Welsh poets from Llywarch Hen to Ceiriog, from Spenser to Tennyson. As it leaves the hills west of the Aran and runs through Llanuwchllyn—"the home of Arthur"—in Merioneth—"as silver clere" to quote Spenser, it enters Bala Lake. Thence it proceeds through the Vale of Edeyrnion and Glyndyfrdwy—"the Glen of the Sacred Waters" and the home of Owain Glyndwr. Soon after passing Llangollen it forms the boundary between Wales and England, and as it enters the Maelor district divides Denbighshire from Flintshire and runs meandering and slow under overhanging trees that dip their branches in its clear waters. Then it becomes an English river for some miles, but taking a bend in a north-westerly direction it flows past Chester and runs its last lap in Welsh territory through Sealand, to form one of the most picturesque estuaries in our island.

The English word Dee is derived from the Latin word "Deva," the goddess of the stream, and Dyfrdwy its Welsh name also comes from the same root. Dyfr = water, and Dwy = holy; the meaning of Dyfrdwy is therefore "Sacred Water."

Like the Dee, the Clwyd has only a short course in Flintshire. It rises in the Silurian hills of Denbighshire in Mynydd Hiraethog, not far from the Merioneth boundary, and bisecting Denbighshire into two fairly equal halves it reaches Flintshire near Bodfari and forms the boundary of the county for some three or four miles. Below Llannerch it becomes a Flintshire river and for the remaining six miles of its course it runs through a district famous in history.

On the right side of the river is Cil-Owain, the place to which Owain Gwynedd retreated after almost annihilating the English army of Henry II at Ewloe Castle. Cil-Owain (Owen's Retreat) does not, however, receive its name from the event, for it appears in Domesday Book. On its left side St Asaph stands on a hill, the home of Christianity from early times, and the resting place of some of the most famous sons of Wales. The Clwyd flows slowly on as if reluctant to leave so beautiful a vale to Rhyd-y-ddeu-Ddwr, "The ford of the two waters," where the Elwy, its largest and last tributary, joins it, then forming a semicircle it passes Rhuddlan Castle, once so famous in Welsh history. The river is tidal beyond the castle, but there are no ships by Rhuddlan Bridge as of old, for the Clwyd is no longer a navigable river and it makes its exit to the sea almost hidden by deep banks of alluvial soil.

The Elwy also rises in the Hiraethog mountains, and after passing Cefn caves enters Flintshire a little above St Asaph. It gave the city its Welsh name (Llanelwy) "the Elwy Church." H. M. Stanley spent his youth on its banks and received his first training in exploring in the neighbouring woods and caves, before he penetrated "Darkest Africa" in quest of Livingstone. Two miles below St Asaph, as we have seen, it joins the Clwyd.

The Wheeler is another tributary of the Clwyd. This

St Asaph

river rises on the east side of the Moel Fammau range above Nannerch and flows into the old line of valley that cleaves the Clwydian range, by Nannerch, thence following the London and North Western Railway towards the Clwyd. By Y Ddol (the meadow) another stream empties itself into it; having risen by Pantasaph it runs through a picturesque glen towards Ysceifiog and the fisheries. On its banks are the ruins of the house where William Edwards (Will Ysceifiog) the poet was born. After passing Afonwen and Maes Mynan, the traditional home of Llywelyn ap Gruffydd, the Wheeler joins the Clwyd at Pont Ruffydd.

But of all the Flintshire rivers perhaps the Alun is the most interesting, especially to a patriotic native, for it travels through the county while the Clwyd and the Dee may almost be described as skirting it. It rises near Llandegla and enters Flintshire a little below the Loggerheads. A short distance down the Leet it sinks and follows a subterranean passage for about two miles, leaving the channel dry as far as the bend near Cilcain stream. Here it issues from the ground and runs as far as where the Cilcain road crosses the Leet; but it is there once more swallowed up in fissures near the bridge, and its channel again remains dry for a distance of two hundred yards north-east of Hesp Alun, where it reappears. There is no visible cavity or hole where it disappears, and the place whence it emerges again is level with its bed. It is a strange phenomenon, as one travels down the Leet, to see the stones in the bed of the river dry, and bridges crossing a waterless course. But when the mountain torrents run, and the heavy rains fall, the subterranean passage is much too small for the volume of water, and it then forms a powerful stream above ground. In the year 1911 it was exceptionally dry and the river travelled underground for seven months.

After reappearing as clear as crystal at Hesp Alun (Dry Alun), the stream runs through a deep wooded

The Leet, Rhydymwyn

valley, where it breaks abruptly through the limestone escarpment and turning almost at right angles to the Wheeler flows south-eastward through the old valley across the coal measures towards Mold, passing on the right Maes Garmon where the so-called "Alleluia Victory" was won. Then it passes Llong and Caergwrle and crosses the boundary of Flintshire, and after skirting

Marford, a detached portion of the county, joins the Dee.

The Terrig rises at Llyn Cyfynwy about three miles from the source of the Alun. For some distance the two rivers run in different directions, but meet again before they leave Flintshire. The Terrig forms the boundary between Nerquis and Tryddyn parishes. At Nant-y-Mynydd several small springs join it from Mynydd Du and Llyn Iorcyn, a small lake on the summit of Mynydd Du, and it meanders through Leeswood Hall park and after running parallel with the Alun for about a mile unites with it not far from Pontblyddyn church.

The Cegidog river runs along the eastern boundary from Ffrith to Cefn-y-Bedd, and joins the Alun as it leaves the county.

Flintshire is famous for its wells, which owe their existence to the Carboniferous Limestone that constitutes its central plain. This rock is porous and the water percolates it till it comes in contact with impermeable shale or clay, where it accumulates and finds its way again to the surface through some of the many fissures. The most famous well is the "Holy Well" of St Winifred, the most copious in Britain. The spring is situated at the foot of the steep hill beneath the town of Holywell; it bubbles up into a beautiful polygonal well covered with a rich arch supported by pillars of fifteenth century date. The well yields on an average 2000 gallons a minute, and during the winter some 3000 gallons. The flow of the spring varies, and is affected by rainfall. The rain that falls on the hills takes about six weeks to reach the well.

Its temperature varies but little and remains almost constant summer and winter from 51·9° to 52° Fahr.

The Holy Well of St Winifred

Another important well is Ffynnon Asa, about a mile from Newmarket in the parish of Cwm. It is the second largest well in Wales, yielding seven tons a minute, and it is said that its volume of water never varies after the longest drought or the heaviest rain; it must consequently rise from a very great depth. Its stream runs towards Diserth where it forms a fine cascade and flows to the sea. Ffynnon Leinw, "the flowing well," in Cilcain parish, was at one time an intermittent spring, flowing at

Hanmer Mere

regular intervals, owing to syphon action, but it has long lost this peculiarity.

The wells were considered sacred by the inhabitants and, as their names suggest, many of them have been dedicated to some saint. Ffynnon Fair (St Mary's Well) in Halkyn, Meliden, and Ysceifiog; St Chad's Well in Hanmer; Ffynnon Mihangel (St Michael's Well) in

Caerwys and Cilcain; Ffynnon Beuno (Beuno's Well) in Tremeirchion and Holywell; Ffynnon Oswallt near Saithffynnon (the seven wells) in the parish of Whitford; Dier's Well near Bodfari; the Goblin's Well, Mold; and Ffynnon Sarah in Nannerch. But the full history of these wells, and the traditions in connection with them, require a larger space than a volume of this kind permits.

Flintshire has but few lakes, and none of any importance. Llyn Helyg, made artificially in the eighteenth century, is the largest, but it is not more than half a mile long and of no great depth. It is situated about four miles from Holywell on the Newmarket road and is surrounded by forest. The Ysceifiog Lake on the road from Caerwys to Mold is also an artificial one, formed by a dam thrown across the valley. It is famous for its trout fishing. Hanmer Mere, in Maelor, occupies a space of about 73 acres and is beautifully situated amid rich woodland scenery.

6. Geology and Soil.

Before speaking of the physical geography of Flintshire, it is necessary to learn somewhat of the geology, as the physical conditions of a county are to a large extent dependent upon its geological structure.

By Geology we mean the study of rocks. The term "rock" is used by geologists without any reference to the hardness or compactness of the material to which the name is applied, loose sand being so described, equally with a hard substance like granite.

Rocks are of two kinds, (*a*) those laid down mostly under water, (*b*) those due to the action of heat.

The first may be compared with sheets of paper laid one over the other. These are called beds, and such beds are usually formed of sand—often containing pebbles, mud, clay, and limestone, or mixtures of these materials. They are laid down horizontally, but may be afterwards tilted or twisted as the result of a movement of the earth's crust, just as we may tilt sheets of paper, folding them into arches and troughs by pressing them at either end.

Rocks, which have been melted under the action of heat and become solid on cooling, are known as igneous rocks. When in the molten state these rocks were poured out at the surface as the lava of volcanoes or were forced into other strata and cooled in the fissures or other places of weakness. Again, much substance was thrown out of volcanoes as volcanic ash and dust, piled up on the sides of the volcano or the country round. Such ashy material arranged itself in process of time into beds, so that at the present day it partakes to some extent of the qualities of the two great groups.

Beds are of great importance to geologists, for by them rocks are classified according to age. If we take two sheets of paper, and lay one on the top of the other on a table we say that the upper one was laid down after the lower. Similarly with two beds, the upper is also the newer, and the newer will remain on the top even after earth movements, save in very exceptional cases. For general purposes, then, we may regard the position of the beds as representative of the age of the rocks, the lower beds being of older formation than the upper.

Earth movements, however, affect the lie of the beds to a large extent. They are often forced upwards from their horizontal position in the form of a "syncline," the newer rock which acts as the cap is worn away, exposing at the surface a rock of older formation, and in time, upon this, other rocks are again laid down horizontally.

After the formation of the beds many changes may occur in them. They may become hardened, pebble-beds being changed into conglomerates, sands into sandstones, muds and clays into mudstones and shales, soft deposits of lime into limestone, and loose volcanic ashes into exceedingly hard rocks. They may also become cracked or fissured, the fissures being often very regular, running in two directions at right angles one to the other. These are known as "joints" and the joints are very important in affecting the physical geography of a district. As the result of great pressure applied sideways, the rocks may be so changed that they may be split into thin layers, which usually, though not necessarily, split along planes standing at right angles to the horizontal. Rocks affected in this way are known as slates.

If we could flatten out all the beds of Wales and arrange them one over the other and bore a shaft through them, we should see them on the sides of the shaft, the newest appearing on the top and the oldest at the bottom. Such a shaft would have a depth of anything between 50,000 and 100,000 feet. The beds are divided into three great groups called the Primary or Palaeozoic,

	Names of Systems	Subdivisions	Characters of Rocks
TERTIARY	**Recent** **Pleistocene**	Metal Age Deposits Neolithic ,, Palaeolithic ,, Glacial ,,	Superficial Deposits
	Pliocene	Cromer Series Weybourne Crag Chillesford and Norwich Crags Red and Walton Crags Coralline Crag	Sands chiefly
	Miocene	Absent from Britain	
	Eocene	Fluviomarine Beds of Hampshire Bagshot Beds London Clay Oldhaven Beds, Woolwich and Reading [Groups Thanet Sands	Clays and Sands chiefly
SECONDARY	**Cretaceous**	Chalk Upper Greensand and Gault Lower Greensand Weald Clay Hastings Sands	Chalk at top Sandstones and Clays below
	Jurassic	Purbeck Beds Portland Beds Kimmeridge Clay Corallian Beds Oxford Clay and Kellaways Rock Cornbrash Forest Marble Great Oolite with Stonesfield Slate Inferior Oolite Lias—Upper, Middle, and Lower	Shales, Sandstones and Oolitic Limestones
	Triassic	Rhaetic Keuper Marls Keuper Sandstone Upper Bunter Sandstone Bunter Pebble Beds Lower Bunter Sandstone	Red Sandstones and Marls, Gypsum and Salt
PRIMARY	**Permian**	Magnesian Limestone and Sandstone Marl Slate Lower Permian Sandstone	Red Sandstones and Magnesian Limestone
	Carboniferous	Coal Measures Millstone Grit Mountain Limestone Basal Carboniferous Rocks	Sandstones, Shales and Coals at top Sandstones in middle Limestone and Shales below
	Devonian	Upper / Middle / Lower — Devonian and Old Red Sandstone	Red Sandstones, Shales, Slates and Limestones
	Silurian	Ludlow Beds Wenlock Beds Llandovery Beds	Sandstones, Shales and Thin Limestones
	Ordovician	Caradoc Beds Llandeilo Beds Arenig Beds	Shales, Slates, Sandstones and Thin Limestones
	Cambrian	Tremadoc Slates Lingula Flags Menevian Beds Harlech Grits and Llanberis Slates	Slates and Sandstones
	Pre-Cambrian	No definite classification yet made	Sandstones, Slates and Volcanic Rocks

Secondary or Mesozoic, and Tertiary or Cainozoic. The lowest of the Primary rocks are the oldest rocks of Britain, the foundation-stones as it were on which the others rest, and these are termed the Pre-Cambrian rocks. The three great groups are divided into minor divisions known as systems.

In the annexed table the names of the great divisions and their systems are given on the left hand side, in the middle the chief sub-divisions of each system, and on the right hand the general characteristics of the rocks.

In earlier ages the arrangement of the land and the sea on the globe was very different from what it is now. The changes were due to movements of the earth's crust, and very largely to movements of compression from the sides, causing folding of the strata of which the crust of the earth is composed. After many and great changes, at a time geologically recent, but still ages before the beginning of history, the district now known as Flintshire rose above the sea for the last time.

Our county lies mainly between two geological disturbances, that of the river Dee on the north-east side, and that of the Vale of Clwyd on the south-west. Within these limits there are no plane or horizontal formations, but the south-western side has been so upheaved as to raise the Silurian as much as 1819 feet high at Moel Fammau, causing the coal measures to drop 1200 feet on the eastern side and, as a consequence of this, all the measures dip in a north-easterly direction.

To the west of the Flintshire hills lies the Vale of Clwyd, composed of alluvium and New Red Sandstone.

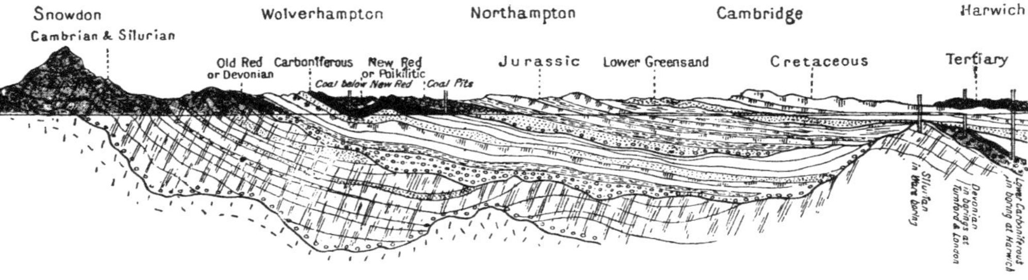

DIAGRAM SECTION FROM SNOWDON TO HARWICH, ABOUT 200 MILES.

The above section is intended to show the order of succession of the rocks in the crust of the earth as it would be seen in a deep cutting nearly E. and W. across England and Wales. It shows also how, in consequence of the folding of the strata and the cutting off of the uplifted parts, old rocks which should be tens of thousands of feet down are found in borings in the South East of England only 1000 feet or so below the surface.

To the east the rich coal measures underlie the ground, which slopes gradually until it touches the alluvial soil of the Dee estuary, while some of the Cheshire Red Sandstone penetrates into a small portion of the north-east of the county.

Though Flint possesses no Pre-Cambrian rocks—the oldest of all—as do the Arenig and the Snowdonian mountains, yet it is famous for its limestone and its chert, and for its productive coal measures.

If we look at the geological map we shall notice that the oldest rock—the Wenlock Shale of the Upper Silurian series—which was once the foundation bed of the other rocks, is now at the top of the mountains. It was formed in an inconceivably remote age and in some great disturbance was thrust upwards for about 3000 feet to form the highest hills in the Clwydian range, namely Moel-y-Parc, Pen-y-Cloddiau, Moel Arthur, Moel Fammau, and Moel Fenlli. It consists principally of roughly cleft shales, with fine-grained sandstones occasionally interstratified, and runs from the Cwm Mountains to the Wheeler valley. It appears again by Maes Mynan and on the driftless slopes of Moel-y-Parc. Its stratified beds can be traced on the southern slopes of Pen-y-Cloddiau. The rock continues through Moel Arthur and Moel Llys-y-Coed, and Bwlch Cilcain, to Moel Fammau, where it can be detected in some old quarries on its southern side. It then proceeds over the boundary to Denbighshire.

There is ample evidence to show that the Carboniferous Limestone of the summit of Halkyn Mountain once formed the bed of a large lake or perhaps the ooze of the ocean itself. The rock still bears testimony to this, for it abounds with fossils and the shells of fish. The Halkyn marble, which is largely used for ornamental purposes, shows abundance of Crinoids, and on the east side of Halkyn the rock consists almost entirely of silicified shells of *Productus*.

The unbroken belt of Carboniferous Limestone extends from Prestatyn to Dafarn Dwyrch, 1½ miles to the south of Llandegla, a distance of 23 miles, where it terminates in a branch of the Bala fault. It varies in width from one mile at Llanferres to 4½ miles between Holywell and Caerwys, and for almost the whole distance it dips in an easterly direction, its outcrop appearing on the west side above the Wenlock Shale, while on the east it is buried under the Lower Coal Measures, being thereby wedged in by two impervious formations. In the northern end of the county—that is, from Prestatyn to a point between Pentre Halkyn and Ysceifiog—the Limestone belt is nine miles in length with an average width of three miles, making the surface limestone area 27 square miles. This area is rich in minerals and contained most of the lead found in the district.

The Limestone, which rises abruptly from the Drift-covered plain near Prestatyn, terminates suddenly as the result of an occurrence of a number of faults of great size, and is then thrown far below the level of the sea. These faults resolve themselves into two principal lines of fracture, one passing from Prestatyn in a southerly direction along the base of Carreg-y-Fran, the other skirting Y Graig Fawr by Diserth waterfall, below Moel Hiraddug and Cwm mountains, and thence up the whole length of the Vale of Clwyd. From its size and its important effects on the physical geography of the district, this ranks as one of the greatest faults of the British Isles.

The ridge of the hill that runs from Pantasaph

Limestone Rocks in the Leet

towards Pen-y-Ball above Holywell consists of hard and thickly-bedded white limestone of great purity. The stone has been quarried in several places and largely exported. In crossing the hills on either side of the Alun, above Rhyd-y-Mwyn, little can be seen, especially on the east side of the river, but terraces of massive grey and white limestone, which is rich in lead in workable quantities.

In Hope Mountain the Carboniferous Limestone is brought to the surface by the great Bala fault in the south and by two almost equally powerful north and south faults on the west, which join at Llanfynydd to form the boundary of the Leeswood coalfield. Here the Limestone takes the form of a massive grey stone mixed with corals and encrinites, probably the continuation of the coral bed at Rhyd-y-Mwyn.

An important subdivision of the Carboniferous Limestone is the Black Limestone, known locally as the "Aberdo," which is of more recent formation than the ordinary Mountain Limestone. Its most westerly exposure is at the Talar Goch mine, but an excellent quality of it is found in great abundance on the Holywell and Halkyn mountains, where it has been extensively worked. Its importance is due to the property which it possesses of forming hydraulic cement when burnt, without being mixed with any other ingredients.

Perhaps the most striking characteristic of the geology of Flintshire is the practical absence of the Millstone Grit. But there is scarcely in any part of the county a more remarkably developed Chert than that which lies above the Carboniferous Limestone in north Flintshire. From the sea near Gronant this bed of Chert runs through Llanasa, Trelogan, Brynford, and Halkyn Hills, east of Rhosesmor; thence, being diverted by faults and covered by the Drift, its course is toward Moel Ffagnallt to appear again in the valley of the Alun above Rhyd-y-Mwyn, having dwindled down to a few feet in thickness. It then passes Pant-y-Buarth, Gwern-y-Mynydd to the east of Moel Findeg, and the Nerquis mountains. It is extensively quarried on the Holywell and Halkyn mountains for the Staffordshire potteries and for use in the grinding mills. The principal chert hills are those at Halkyn Mountain, near Henblas, Moel Gaer, Bryn Gwyn, and Nerquis Mountain. In the Trelogan mines it extends from the surface vertically to the depth of 175 yards. It consists of almost pure silica with the exception of a few thin shale bands.

The red Bunter sandstone does not enter to a great extent into the county. It is present at Broughton and Kinnerton on the east. The eastern half of Maelor consists of the red marl so prevalent in Cheshire, but the western part (the Overton district) is formed of Bunter.

The sands and Boulder Clay in the county belong to the glacial period. This clay is formed in irregular pipes or pockets in the limestone on the line of fault. White clay (marl) found in one of these fissures near Caerwys was used largely in the manufacture of tobacco pipes. It was quarried on Moel Ffagnallt, Moel-y-Crio, Glan Alun mine (near the Loggerheads) as well as at other places.

Superior bricks can be made of Boulder Clay when free from limestone, but it invariably contains limestone pebbles and in some instances a heavy deposit, which renders it worthless for the manufacture of brick or any kind of earthenware. The presence of this limestone in the clay proves that the Drift traversed the county from the west and south-west. The clay was used for glass-making, while the white cherty sandstone (which contains 80 per cent. of silica) was ground, and employed in the manufacture of china and earthenware. When it is mixed with common potter's clay it forms a stoneware of a quality superior even to that of the finest Staffordshire ware. The clay is derived mostly from the upper beds of the limestone and in many instances it has been brought to its present position through the action of an underground stream. It varies in composition in different parts of the county according to the rocks over which it travels; gravelly if it is a superstratum of sandstone, clayey if of slate, red if of New Red Sandstone, yellow if of coal-measures and Wenlock shale. Along the coast the Boulder Clay is of a purple colour and contains numerous stones 18 to 20 cubic feet in bulk from the Lake District. At Kelsterton the Boulder Clay is 16 feet thick and rests on the sandstone.

The superficial deposits that form the soil and surface of the county consist of three subdivisions in the Boulder Clay—sand and gravel of the glacial period; the tufa; and the alluvium, of recent period, found at the estuaries of the Clwyd and Dee. Peat beds also occur in small deposits at different parts of the county on the mountain as well as in the marshes, but peat is present in large quantities in Fenn's Moss on the south-east boundary of Maelor, and is so soft that it has to be cast in moulds to harden. It is then sold as fuel and is very flammable.

There is at Pwll Gwyn and Afonwen a very large deposit of calcareous tufa (white marl or soft limestone), covering about 15 acres, and varying in thickness from a few inches to 45 feet, and wherever tested it is found to be resting on Drift sand and gravel. This tufa contains

96 to 98 per cent. of carbonate of lime and is now being extensively used in the manufacture of Portland cement. As the limestone in the immediate district is hard and in well-defined and regular beds and rather highly coloured by oxide of iron it is quite evident that this recent deposit is not a native of this place. About two miles to the north-east of Afonwen and 300 feet higher, there is a large tract of country known as "Ffrith y Garreg Wen" (the Moor of the White Stone). This tract contains the upper series of the Carboniferous Limestone free of oxide of iron. The natural fall or wash of this tract is to the east, where a ravine known as "Trefraith" has been formed, running in a southerly direction and expanding into a chasm, north of Caerwys, its stream forming a tributary of the Wheeler at Afonwen. The natural conclusion is that the tufa or marl, having "Ffrith y Garreg Wen" as the gathering ground, was washed down and deposited in its present position.

The geological map shows that the coal measures underlie a large part of the county, for the district east of the line drawn from Nerquis Mountain skirting Halkyn to Gronant is in the coal area. This productive coal-field may be divided into three divisions: northern area—Mostyn to Flint; eastern area—Northop, Buckley, Hawarden, and Queensferry; southern area—Mold, Nerquis, Tryddyn, and Leeswood.

In the eastern and southern area the seams are correlated. The northern area is somewhat different, for it contains 15 seams measuring from a few inches to 15 feet, making a total of 78 feet. Of these, nine are workable seams, forming an aggregate of over 60 feet. Only a very small percentage of this area has been worked, leaving a large tract of virgin coalfield on the estuary of the Dee. Although the Mostyn and Bychton coal seams were worked as early as the thirteenth century, the only collieries now existing here are the Point of Ayr in the extreme north and the Bettisfield Colliery at Bagillt. The eastern area is being worked at Northop Hall, Buckley, and Hawarden.

The chief collieries of the southern area are those at Mold, Nerquis, Tryddyn, and Leeswood (which is of peculiar geological formation). An extraordinary deposit of Cannel coal was discovered at Leeswood in 1858 and worked extensively for many years. This coal had excellent gas-producing qualities and was chiefly used for the manufacture of paraffin oil by distillation at a low red heat, yielding 80 gallons of crude oil to the ton.

In 1865 some 150,000 tons of Cannel coal were raised and over a thousand retorts were erected for distilling the oil, but it is now practically exhausted.

7. Natural History.

Although Flintshire is the smallest of the Welsh counties it possesses a rich and varied fauna and flora, and whether regarded numerically, or from the standpoint of rare and interesting species, it will afford favourable ccmparison with many larger counties.

An examination of our Flintshire plants shows that a

large number of the species are of the so-called English type; plants, that is, having their headquarters in southern England. This southern element, indeed, is quite a feature of the county flora. Among numerous examples of this kind are the columbine, greater celandine, violet, maple, and dogwood. A number of these English species rapidly disappear after reaching Flintshire, and in one or two cases, e.g. wild madder (*Rubia peregrina*), this forms the northern limit of the plant's range. The Atlantic or Western coast type is also represented, and includes such plants as the wall pennywort (*Cotyledon umbilicus*), the sea spurge, and Portland spurge. The dainty little ivy-leaved bell-flower (*Wahlenbergia hederacea*), which occurs in one Flint station, is also to be classed in this division. The butterwort (*Pinguicula*), crowberry (*Empetrum nigrum*), and mountain pansy (*Viola lutea*) are examples of the Arctic type which descend into Flintshire. Plants specially characteristic of the Scottish Highlands

Cowberry (*Vaccinium Vitis-Idaea*)

are almost entirely lacking, but the cowberry (*Vaccinium Vitis-Idaea*)—which grows on Gwern mountain—supplies an example of this type.

Typical alpine plants are scarcely represented with us, as our greatest elevation, Moel Fammau, is only 1819 feet above sea level. A tiny moss (*Bryum alpinum*) and several allied plants which occur at the Point of Ayr are, however, essentially alpine species. Probably the explanation of their occurrence in this unusual situation, practically at sea level, must be sought in the glacial period of long ago.

The great variety of plant life for so relatively small an area is largely explained by the diversity of the underlying rocks, which of course determine the nature of the soil. The flora of the Clwydian range, which consists of Wenlock shale, affords comparatively little variety and is essentially of a drought-resisting (xerophilous) character. Bracken, bilberry, ling, and dwarf furze (*Ulex Gallii*) are typical species along the greater part of this range, and in the late summer the blossoms of the two latter colour the hillside with glowing sheets of purple and gold.

As we pass on to the limestone formation the vegetation undergoes a remarkable change, and we now meet with a noteworthy variety and luxuriance of plant life. The rock rose, dogwood, spindle-tree, salad burnet (*Poterium Sanguisorba*), and yew, are just a few of the many lime-loving plants which may generally be seen on such soil. These limestone districts are the haunt of many rare and interesting forms of plant and animal life, and so we find that such places as the Leet, Diserth, and

Halkyn Mountain are deservedly favourite localities for the naturalist.

It is interesting to observe that certain species of plants are often restricted to some particular type of geological formation. About Cwm Mountain (Wenlock shale) and the Clwydian range generally, the foxglove ("Bysedd-Cochion") is frequently seen. As soon as one crosses on to the limestone it immediately disappears. This curious influence is well shown in the course of a few minutes walk from Cwm Mountain to Moel Hiraddug.

Sea Spurge (*Euphorbia Paralias*)
Flintshire sand-hills

The influence of man on vegetation, though often unconsciously or accidentally exerted, is nevertheless frequently very marked, and our own county affords a number of cases in point. There are few districts in Flintshire where one may not encounter the greater celandine ("Llym y Llygad"), which is now thoroughly naturalised. The wormwood and lucerne are further instances of this kind—"denizens" as they are termed, that is, plants which were originally cultivated for purposes of medicine, food, or ornament, but which have escaped and now appear more or less native. The mimulus or monkey flower, a native of South America, is quite naturalised along the banks of the Clwyd. The ivy-leaved toadflax (*Linaria Cymbalaria*), which is common on walls in many parts of the county, was accidentally introduced into Britain from Italy. The whitlow pepperwort (*Lepidium Draba*), although not a native plant of our islands, is now thoroughly established in the county and is quite a common species in various places along the Dee estuary. On the Dee embankment for a distance of half a mile or so north-west of the Bettisfield colliery it is the predominant plant. This alien is a native of central and southern Europe, and it appears to have been introduced to Britain with the bedding straw of invalid troops disembarked at Ramsgate from the Walcheren Expedition. Owing to the absence of large industrial centres, however, alien plants are not so well represented in Flintshire as in Lancashire.

It is worthy of note that the average annual rainfall along our Flintshire coast is very low, and this fact is not without influence on plant life.

So far as is known Flintshire possesses no species of plant which is peculiar to the county, but on the other hand the flora includes many rarities. Among these we may mention the round-leaved winter-green (*Pyrola rotundifolia*). This handsome plant, which occurs in one station only, is found in no other Welsh county, and is also noteworthy as being the only representative of the genus in the Principality.

The Vale of Clwyd and much of Flintshire and North Wales were formerly a vast forest. Traces of this ancient jungle may still be seen in the "submerged forest" at

Rhyl. For our knowledge of the old-time inhabitants of this wild region we must turn to certain limestone caves at Tremeirchion and Y Gop, by Newmarket. These caves, which were evidently used by various wild animals for shelter over a lengthy period of time, are remarkable for their remains, bones and teeth being found here in extraordinary abundance. No less than sixteen different species of mammals have been identified in the deposits of the two caves at Tremeirchion. The list is as follows: Cave lion, spotted hyena, woolly rhinoceros, mammoth, Irish elk, wild cat, wolf, fox, brown bear, badger, wild boar, Celtic short-horn, red deer, roe deer, reindeer, and horse.

In a hasty survey of our Flintshire birds we notice that several shore species do not breed in the county owing to the absence of suitable cliffs and rocks along the coast. A colony of lesser terns make their home in one shore station every year, and it is satisfactory to know that their eggs and young are carefully protected. Several gulleries occur in secluded situations along the Clwydian range, whither the black-headed gull retires every spring. The raven is occasionally met with in Flintshire but does not now breed in the county. The hawfinch is now not uncommon and appears to be increasing. The large expanses of salt marsh along the Dee estuary are excellent localities for studying winter visitors and have long afforded a favourite resort for the snow-bunting and many species of geese and ducks.

The first British example of the nutcracker crow was recorded from this county. This was shot at Mostyn in 1753 and was described and figured by Pennant, our Flintshire naturalist, in his *British Zoology*. Of course this can only be regarded as a very rare stray visitor to these islands, and the spoonbill, which has also been seen in the county, must be placed in the same category.

The viper is perhaps the most noteworthy Flintshire reptile and is not uncommon, especially amongst the heather on the dry sunny slopes of the Clwydian range. This is our only poisonous British snake and it may be readily distinguished from the harmless grass-snake by the prominent black zigzag line along the back. The natter-jack toad frequents the coast at Rhyl, but it seems to be more plentiful on the opposite side of the Clwyd between Foryd and Pensarn in Denbighshire. This is a distinctly local amphibian. It is smaller than the common toad and is easily recognised by the yellowish line running down the back.

8. Coastal Gains and Losses.

An interesting chapter in our history is afforded us in the story of the battle between land and sea. This battle has lasted from the time when the myriads of shells on the Halkyn mountains were inhabited up to the present time.

At one period the sea gained the mastery over the land, to be checked again by some upheaval of the earth. The Vale of Clwyd once formed a bay and the waves of the ocean lashed against the Hiraddug rocks. When the

Iberian dug his tumulus and built his caer on the hills, the land ran out for miles to the west, where now large ships sail. The Dee in Roman times was fordable at many points and the Mersey it is said was a tributary of it and joined it at Chester.

In the fifth century there was a great depression along the whole of the North Wales coast, and the sea once more regained a large tract of land. Reference to this overflowing of the land is made by Welsh bards. Taliesin, who lived in the fifth century, said:—

> "Seithenin sav di allan, ac edrych
> Uirde varanes mor maes Gwytneu rhytoes."

> i.e. "Seithenin, stand, look, and see
> How Gwyddno's land is deluged by the sea."

The Lowland hundred, Cantre'r Gwaelod—North and West Wales—had as its prince Gwyddno Garanhir—i.e. Gwyddno Longshanks. His territory is to-day covered by the waters of Cardigan Bay. At that time he owned a fishing weir at the estuary of the Conway; hence his connection with that part of Flintshire, which is now also covered by the sea. This fertile tract was called Morva Rhianedd. When the present coast-line is compared with that of the fifth century it is seen that the ancient Flintshire must have been nearly twice as large as at present.

At a very low ebb tide the remains of a Sarn or causeway called "Muriau" (the walls) are still to be seen about four miles below Llandrillo church. This corresponds to the ancient place-name Pen Sarn, that is "the water road end," near Abergele. A stone embedded in the Abergele churchyard wall also alludes to the submerged land:—

"Yma mae'n gorwedd, yn mynwent Mihangel
Ddyn oedd a'i annedd
Dair milltir i'r Gogledd."

"Here lieth a man, in St Michael's churchyard, whose dwelling-place was three miles to the north." This, if correct, places his home at least two miles out to sea.

Bishop Trevor of St Asaph (1395–1410) complained that the sea had eaten up his land at Gronant-is-y-Mor, between Mostyn and Prestatyn, and made an application for reduction of rent.

There are instances of sea erosion in later times. Prestatyn Castle, now a mile from the sea, stood originally about three miles distant from it. This distance compares favourably with that of the Abergele stone, while Sarn Badarn, mid-way between Prestatyn and Gronant, corresponds with that of Pen Sarn, near Abergele, where also at low ebb the remains of a causeway were once visible. Moreover, tradition has it that the inhabitants of Puffin Isle, in times gone by, used to come and record their votes in Flintshire. Maps and deeds as late as the time of Queen Elizabeth are extant which prove that these lands now submerged extended three and more miles northwards of the existing waterline. Four persons over 84 years of age now living in Flintshire have seen corn cut where to-day heavily-laden vessels sail to and from the adjacent quarries.

The Mostyn and Flint marshes, which in 1870–1880

Prestatyn from the Hills

were pastures intersected with tidal rills, are now totally devoid of vegetation and are of no value for agricultural purposes. Rapid sea wear is manifested on a large scale. Between 1871 and 1898, 100 yards disappeared at one point. More than a mile of the Prestatyn coast is subject to erosion at the rate of about a yard per annum. There are traces in this locality of old water-tracks and fords along which rivers, now dried up, or inlets from the sea once flowed. East of Rhyl there are farms called Rhyd-Orddwy "the ford of the Ordovices," and Rhyd (a ford), names which possibly have reference to another stream.

Erosion has been very rapid of recent years. All the sand-banks in the parishes of Rhuddlan, Diserth, and Meliden, which are the only protection from the sea, are gradually disappearing, and according to one report the districts on the sea coast are likely to suffer seriously unless some protective work of substantial character is carried out soon.

It is estimated that about 250 acres have been lost between Rhyl and Prestatyn since 1794. The wearing away of the sand-hills will eventually render the lowest lying portion of Rhyl subject to floods. Owing to heavy rains occurring when the water-courses were tide-locked, a large portion of the town was flooded some forty years ago and the sea brought a great whale to the bottom of High street. The town is built on the landward side of the sand-hill range, three feet below the high-water mark of a 21-feet tide. More than half the town is on land below high-water mark. Although a slight gradient exists along the ridge up to St Asaph there is nevertheless a fall inland, in some places as much as nine feet, on the east side and west side of the vale. Hence arose the necessity for erecting the cops (embankments) on both sides of the river from the junction of the Clwyd and the Elwy to the Foryd, a distance of about four miles. Long ago the waves used to wash over the defences on the west of the Clwyd and the whole marsh was flooded. Some of the

The Sand-hills, Rhyl

oldest inhabitants remember boats being used to carry provisions from Rhuddlan to the marsh-dwellers, who secured themselves against the floods by resorting to their upper storeys.

When the Court Leet for the Manor of Englefield met in 1912 it was reported that considerable sea erosion was taking place outside the Rhyl boundary. The sea was within five yards of land belonging to the Crown, and the golf links were threatened. To check the advance of the sea the Railway Company have built a magnificent wall west of the Clwyd as well as between Holywell and Mostyn.

In old maps the estuary of the Dee is shown as an open arm of the sea. Long ago the river was tidal as far as Chester. Vessels were brought up to the walls of the city, which was then the port for Ireland, but owing to continual changing and silting up of the channel of the river a new quay was constructed at or near Shotwick, about six miles from Chester, and in 1560 a collection was made in all churches throughout the kingdom towards the cost. In the year 1732 an Act of Parliament was procured, enabling the mayor and citizens of Chester to recover and preserve the navigation of the Dee. In the year 1740 "The River Dee Company" was formed. It received, as a recompense for recovering and preserving the navigation, a grant of all "the white sands" within the estuary of the Dee, from the walls of Chester to the extremity of the Wirral on the Cheshire side, and to the Point of Ayr on that side of Flintshire. This company purchased 600 acres of waste marsh land, through which they cut a new channel for the Dee, and several embankments were made in the years 1754–1790, when 3100 acres were reclaimed. The whole of this redeemed tract has been formed into the parishes of Sealand and East and West Saltney. There are still between 1000 and 2000 acres of unenclosed marshes on the estuary of the Dee, the principal portion of which is in the vicinity of the towns of Flint and Holywell, and consists of land of the richest quality. About the year 1870 a bank was constructed from Burton Point to Connah's Quay, but it was broken through by the tide before its completion, and the marsh behind it, nearly two miles in extent, remains still unenclosed. There is an important scheme now under

The Sands of Dee

consideration to undertake the navigation of the Dee by cutting a straight navigable course from Mostyn to Connah's Quay, at an estimated cost of £75,000. Although the waste lands are still of considerable extent, more particularly on the hills of Buckley and Halkyn, yet they have been greatly lessened by enclosures. Some of the chief enclosures are that of Saltney Marsh, containing

2200 acres, under an Act passed in 1778; that of Hope, comprising 3500 acres, under an Act obtained in 1791; that of Mold, comprising about 4000 acres, under an Act passed in 1792; that of Cilcain, containing 2400 acres, under an Act passed in 1793; and those of Ysceifiog, Nannerch, and Whitford, comprising about 3500 acres. In the parish of Llanasa, by an Act passed in 1811, 1600 acres of peculiarly rich land have been enclosed, of which 1200 were recovered from the sea by an embankment at an expense of £4000, defrayed by the freeholders. In 1807 the proprietors within the franchise of Rhuddlan obtained an Act for the enclosure of the rich tract called the Rhuddlan Marsh, which saved about 27,000 acres of rich sandy loam. Some 500 acres were allotted by an Act of Parliament to be sold towards defraying the expenses incurred in making the embankment, which was estimated to cost, together with the drainage, as much as £13,500.

9. Climate and Rainfall.

The climate of a country depends upon various factors, all mutually interacting; upon the latitude, the temperature, the direction and strength of the winds, the rainfall, the character of the soil, and the proximity of the district to the sea.

The difference in the climates of the world depends mainly upon latitude, but a scarcely less important factor is proximity to the sea. Great Britain possesses, by reason of its position, a temperate insular climate, but its average annual temperature is much higher than could be expected from its latitude. The prevalent south-westerly winds cause a drift of the surface waters of the Atlantic towards our shores, and this warm-water current is the chief cause of the mildness of our winters. Most of our weather comes to us from the Atlantic. It would be impossible here within the limits of a short chapter to discuss fully the causes which affect or control weather changes. It must suffice to say that the conditions are in the main either cyclonic or anticyclonic, which terms may be best explained, perhaps, by comparing the air currents to a stream of water. In a stream a chain of eddies may often be seen fringing the more steadily-moving central waters. Regarding the general north-easterly-moving air from the Atlantic as such a stream, a chain of eddies may be developed in a belt parallel with its general direction. This belt of eddies, or cyclones as they are termed, tends to shift its position, sometimes passing over our islands, sometimes to the north and sometimes to the south of them, and it is to this shifting that most of the weather changes are due. Cyclonic conditions are associated with a greater or less amount of atmospheric disturbance, anticyclonic with calms.

The prevalent Atlantic winds largely affect our island in another way, namely in its rainfall. The air, heavily laden with moisture from its passage over the ocean, meets with elevated land-tracts directly it reaches our shores—the moorland of Cornwall and Devon, the Welsh mountains, or the fells of Cumberland and Westmorland—

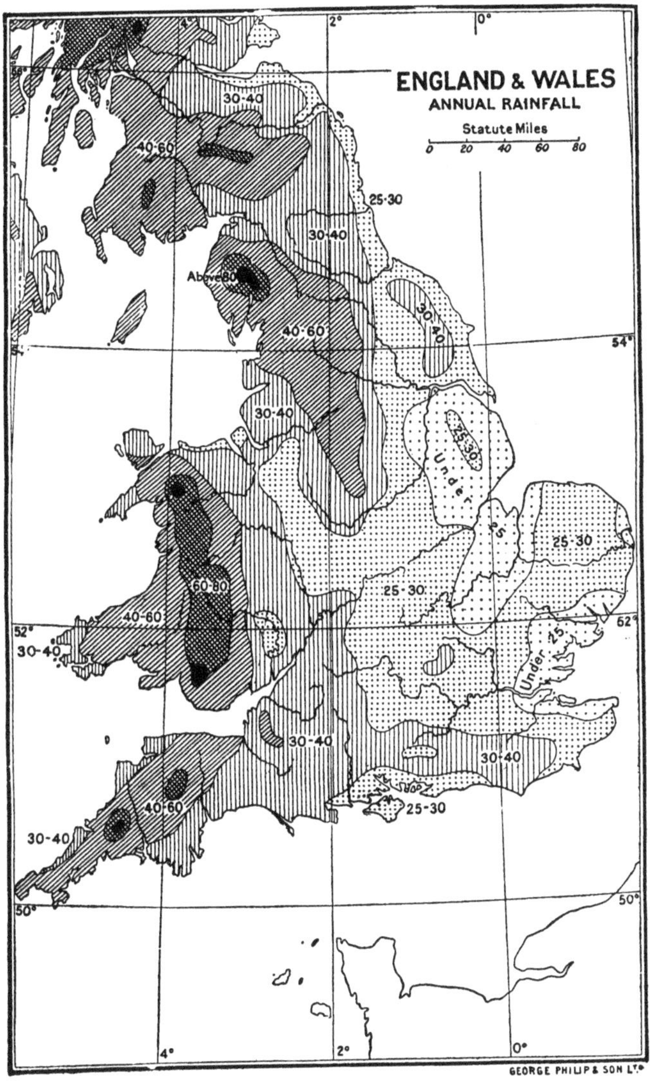

(The figures give the approximate annual rainfall in inches)

and blowing up the rising land-surface, parts with this moisture as rain. To how great an extent this occurs is best seen by reference to the map of the annual rainfall here given, where it will at once be noticed that the heaviest fall is on the high mountains in the west, and that it decreases with remarkable regularity until the least fall is reached on our eastern shores. Thus in 1908 the maximum rainfall for the year occurred at Llyn Llydaw in the Snowdon district, where 237 inches of rain fell; and the lowest was at Bourn in Lincolnshire, with a record of about 15 inches. The Welsh mountains may not inaptly be compared to an umbrella, sheltering the country farther eastward from the rain.

Flintshire is on the same latitude as Irkutsk in Siberia and Bering Sea on the one hand, and Labrador, Hudson Bay, and Lake Winnipeg on the other, a fact which conclusively shows that latitude and temperature do not always correspond. Isothermal lines are used to mark out places that have the same temperature, and if we follow the isothermal line of Flintshire in July we shall see that places within a few degrees of the Arctic Circle and other places only a few degrees outside the Tropics, enjoy the same average temperature as Hawarden or St Asaph. Our summer is as cool as that of Finland or the north-west of Canada. But we should find the isotherms of these places very different in winter.

Flintshire has a temperate and fairly equable climate, as the limestone which forms the greater part of the soil quickly drains all moisture; and it has a dry, bracing, and invigorating air tempered by sea-breezes. The climate along the coast is as genial as that of the south coast of England. This is due to the protection afforded by Ireland and the Snowdonian range. The mean annual temperature at Rhyl in the year 1910 was registered at at 50·2°, which is much the same as that of Torquay and other southern health-resorts.

The most prevalent wind is the south-west, which blows on an average twenty days a month. The east wind has lost its sting to a very great extent before it reaches our county.

Rainfall plays an important part in determining the climate of a place. The average yearly fall for the whole of England and Wales is about thirty-five inches, for Scotland about forty. The average rainfall along the Dee estuary is low, the reason for this being that the prevalent winds that blow from the sea laden with moisture meet no obstacles till they reach the Halkyn mountains and the Cheshire hills. The return for this district in 1875 to 1881 (1876 missing), during the five years of the six years given in it, shows the lowest rainfall in Wales, and the rainfall at Rhyl in 1910 was only 25·89 inches, compared with 31·35 inches at Eastbourne and 58·02 inches at Torquay. At Mold and Halkyn there is, of course, a heavier rainfall than along the coast. From records kept at Soughton Hall, 425 feet above sea level, between Northop and Mold, we find that the average rainfall for the last thirty years was 32·57 inches. At Tremeirchion the average yearly rainfall from 1893 to 1910 was 27·95 inches, and the average number of rainy days was 159.

There is a marked difference between the rainfall on

the two sides of the Vale of Clwyd. The hills on the western side seem to hold the clouds back, especially when there is a light south-westerly wind, and as the air is moist on such occasions, it rains heavily on the Hiraethog mountains for a long spell, while farmers on the east side of the valley enjoy fine weather.

Another point of interest in this district is the formation of very heavy fogs or mists across the whole valley whenever there is an east or south-easterly wind, due probably to the fact that the district is warmer and more sheltered than its neighbours.

From records kept during the year 1911 at St Beuno's College, Tremeirchion, on the eastern side of the Vale of Clwyd, we find that the highest temperature in the shade was 90·2° Fahr. on August 13, and the lowest 26·4° on February 2.

In 1908, there were recorded 1674 hours of bright sunshine on the coast. Comparing this with other places it will be noticed that our coast gets a good share of sunshine, Manchester recording 991 hours, Birmingham 1140 hours, Scarborough 1378 hours, and Buxton 1298 hours during the same period. In 1911, Rhyl, from April to September (185 days) had only five sunless days. The number of hours of sunshine on the Clwydian hills in the same year was 1391, and as the recording instrument is placed on the side of a hill a couple of hours is often lost in the morning, but in the years 1908, 1909, 1910, and 1911 it recorded 1186, 1207, 934, and 1391 hours respectively, making an annual average of 1180 hours for 1908–1911 inclusive.

10. People. Race. Dialect.

We have no written records of the first men who lived in our county long ages ago. Writing was an unknown art, and records, even if they had existed, could not have survived to come down to us. We therefore speak of this period as Prehistoric, the time when the people of the past were unable themselves to record their story. Yet though these sources of information are closed to us, we are able from the relics they have left behind them, the implements and the weapons they used, the bones of the animals they fed upon, the structures they erected, to form a fairly clear idea of these early people.

The Prehistoric period, vast in its extent, has for convenience sake been further subdivided. At first the metals were unknown, or at least unused, and this period is spoken of as the Stone Age, for it was of flint and other stones that the weapons and domestic implements were mainly fashioned. Later, man learnt how to get the easily-worked ores of tin and copper from the rocks and by their admixture to form bronze. From this, beautiful weapons and other articles were made. This period was called the Bronze Age. Doubtless the ores of iron had long been known, but how to smelt them was another matter. At length the method was discovered, and mankind was in possession of hard metal implements having great advantages for all purposes over the flint and bronze weapons previously employed. Thus the Iron Age began and the inhabitants of our island had arrived at this stage of civilisation when the Romans came to Britain.

The earliest inhabitants of Flintshire, the people of the Palaeolithic Age or Old Stone Age, have left no traces. They lived in caves or on the banks of the rivers, when the mammoth, reindeer, and hyena, whose bones have been found in Tremeirchion and Cefn caves, roamed over our country. The people made leaf-shaped roughly-flaked flint weapons, which were never ground or polished;

Fair Day at Mold

they cultivated no plants and tamed no animals; neither did they build houses, graves, or monuments. They disappeared, and later—but how much later we know not—the district, it is believed, was peopled by an Iberian race from south-western Europe, a race owning flocks and herds, and possessing a knowledge of many arts and crafts—such as spinning and weaving and the making of pottery and of dug-out canoes—but having at first no acquaintance with the use of metals. These were the people of the Neolithic Period, of the same stock as the Silures of South Wales and probably dark-haired and black-eyed, round-headed and short of stature. Their descendants may still be seen amongst the colliers at Bagillt, or on a fair day at Mold, and it is quite possible that their breed of domestic animals may be represented upon Flintshire farms to-day.

The Iberians, it is thought, were conquered and driven westward by the very different Goidels or Gaels, a powerful Celtic race, tall, fair, long-headed, much further advanced in art and craft, and to some extent users of bronze for tools and ornaments. It is thought by some authorities that it was they who set up the stone circles, avenues, and menhirs. Many of the people of Scotland and Ireland are their descendants, and their language is still spoken in the Highlands and the western isles of Scotland, in the south and west of Ireland and in the Isle of Man. The Gaels again were followed and conquered in the fourth century before Christ by the Brythons, another Celtic race, who gave their name to our islands. They took possession of Wales, and of Scotland as far as the Highlands, but they do not appear to have crossed to Ireland. They were to a great extent users of bronze, but they also worked in iron and were the first people of the Iron Age to make their home in our country. It is probable that they built most of the forts and caerau on our hills.

In historic times advancing peoples spread over what is now Flintshire. The county was under Roman rule for 400 years. The Teutonic tribes followed them, and for another 400 years they partially conquered and settled in parts of the county. The names Prestatyn, Mertyn, Mostyn and others still remind us of them. The Danes settled on the Wirral peninsula, but as Chester was then in the possession of the Saxon and as there were no rich monasteries to tempt the pirates over the estuary, few only lived along our coast. Point of Ayr, Gwaenysgor, and Aber Kinsey near Rhyl are their old abodes.

Then came the Normans, whose castles from Chester to Rhuddlan are proof of the thorough and lasting character of the conquest.

Conquest gave place to peaceful settlement in the Middle Ages. Then many Irish found homes in this county, though they have left behind no place-names as in Anglesey. Even to the present day this immigration has continued into such places as St Asaph, Holywell, and Flint.

During the last two centuries English folk from Staffordshire have made their homes in Buckley and the district, on account possibly of the identity of trade. They brought their own dialect with them, which is probably the origin of the "Buckley brogue" now predominant in a circle which includes Buckley, Hawarden, Connah's Quay, Broughton, Kinnerton, Pen-y-Mynydd, Pen-y-Ffordd, and intervening hamlets.

Along Halkyn mountain live the descendants of Cornishmen and Derbyshire immigrants who once worked in the lead mines. These retain their Cornish and English surnames, but have adopted the Welsh language instead of their own.

In the Census Table of 1911, out of 75,931 persons from three years of age upwards, 5722 were returned as speaking Welsh only, 31,568 as speaking Welsh and English, and 38,544 as speaking English only.

Around Saltney, Hawarden, Buckley, and Hope, and along the railway track from Sandycroft to Flint, the great bulk of the population speak English only. Along the coast from Bagillt to Rhyl, at Mostyn, Ffynnon Groew, Gwespyr, Gronant, Gwaenysgor, Meliden, Newmarket, and Diserth; around Rhuddlan, Rhuallt, and Bodfari; and in more inland towns and villages of the county, the bulk of the population speak both Welsh and English.

The Welsh spoken in Flintshire is not unlike the Welsh one hears in the Dee valley, though it has a few peculiarities of construction.

English is spoken now in Maelor Saesneg but the Welsh place-names of fields and streets remind us of a time when Welsh was the language of this district.

11. Agriculture.

The area of Flintshire is approximately 164,000 acres, or, excluding water, 162,445 acres. Of these, 39,662 are under cultivation; 86,975 acres yield a good hay harvest, while mountain and heath land occupies 11,275 acres. Plantation and forest cover an area of 7470 acres.

The Board of Agriculture returns show that Flintshire, although a small county of only 257 square miles, ranks high in comparison with other Welsh counties as regards agriculture. Taking the figures of the year 1911 we find that oats, now the chief cereal, occupy 10,475 acres, yielding 28·73 bushels; followed by barley covering 4125 acres and yielding 28·69 bushels to the acre. Barley, however, is no longer the principal grain used, as in the last century. Much more than was needed for local consumption was then grown, and largely attended

Flintshire Cattle

markets were held weekly at Holywell and Mold. Wheat and barley were exported in large quantities to Liverpool.

In common with all the other counties there is much less land under the plough now than in former years. This is chiefly owing to the reduction in the price of wheat, and the increase in the cost of production. The tilling of the soil has become secondary to the raising of stock. During the last ten years (1901–1911) 158,000 acres in Wales have been converted from arable land to pasture.

The live stock returns for 1911 show 6393 horses for this county—the smallest number in any Welsh county except Merionethshire, and in England except Rutlandshire. The return of cattle was 39,963. Four English counties as well as Anglesey and Merionethshire are behind Flintshire in this respect. It has 102,923 sheep, the smallest number in any Welsh county except Anglesey, and it has 20,328 pigs, being in this respect the sixth richest county in Wales. If these animals were equally distributed among the population, there would be six sheep between five persons, one head of cattle between two persons, one pig between every four persons, and one horse for every twelve persons.

The number of persons engaged in agriculture forms but a small proportion of the population. According to the 1911 census only 4712 are returned as farmers and farm labourers. More than that number find employment in one of the chief industries of the county.

The higher portion of the county is mountainous and is not adapted for agriculture. It is covered to a large extent with heather and coarse grass. The soil is composed of inferior loam and is in many parts thin. The limestone predominates in some parts and forms large tracts of rock. There are certain parts as yet uncultivated and suitable for agriculture, and the Small Holdings Act will undoubtedly encourage more to enclose and cultivate the land. Several farms and houses on the Common are copyhold—that is, the owner has no title deeds to his farm or house, but a record is kept by the lord of the manor. The Common that runs from Pantasaph to Rhosesmor is free for grazing, and small mountain sheep thrive well on it. The sheep produce little wool and their fleece is coarse, but as they find their own food in the choice herbage of the mountain side they yield mutton of delicious flavour.

A Flintshire Farm House

Years ago the hardy black cattle, which were left out the whole year round, were to be seen among these hills. There are but few to be seen now and other cattle have replaced them. The predominating breeds are the Shorthorn, Argyllshire, and a few Devonshire, as well as many cross-breeds. Although the feed is scarce, the butter from this district is of fine quality and good flavour.

There are no sheep-walks belonging to the farms of the county as in Carnarvonshire. The sheep are brought from Merionethshire and other counties and fattened on the land where wheat used to grow for local consumption or for the English markets.

The size of the farms varies from two acres to four hundred acres. The Mostyn estate, for example, includes 52 farms of 2 acres and under 10 acres, 24 farms of 10 acres and under 20 acres, 26 farms of 20 acres and under 50 acres, 23 farms of 50 acres and under 100 acres, and 16 farms of 100 acres and under 400 acres.

Although most of the farmers keep poultry, there are no poultry farms as such. Beekeeping is only a casual occupation, and the same may be said of fruit culture or horticulture, except along the estuary of the Dee. The county is well wooded and the dingles abound with large oaks, beech, and other trees.

Maelor district has a good deal of arable land laid down into pasture. There is no fallowing. The making of cheese and fattening of pigs receive much attention. The dairy is the mainstay of the district. The average sized farm here measures about 100 acres and is rented at from one to two pounds per acre. Few fields are ploughed for roots and straw. The soil is clay and greatly improved by drainage. There are no sheep-walks and no copyhold lands in the district. The farm buildings are usually of brick, made in the locality.

As limestone is plentiful in the county almost every farm once had its own lime-kiln, which was called "aelwyd" (hearth). It was important to maintain a steady current of air in the old sod kilns, so the "aelwyd" was built on a spot convenient to catch the prevalent winds. A place near Holywell is still called "Saith Aelwyd" for seven kilns once stood there. Hardly any lime is now used for agricultural purposes and the few kilns that burn lime supply the building trade only. Another fact significant of the change in agriculture is the almost entire absence of the small mills that were once so numerous. Every rill had its one or two mills, but now all the streamlets run idly into the sea, for wheat is no longer sold at 21*s.* per bushel.

The eastern district has a different standard of weights and measures from that used in the western portion. The English square rood was the fourth part of an acre, whereas the Welsh rood was 64 square yards. The yard by which the Saxon sold his goods was 36 inches, but the Welsh yard was 40 inches, which suited the Saxon well when he bought from the Cymro. The Saxon farmer hired his servant by the month, the Welshman engaged his "gwas" by the "tymhor" (or six months) or for twelve months. The Saxon sold his wheat in measure, and a measure is 75 lbs., whereas the Welshman used the "hobbet." A hobbet of wheat is 168 lbs., a hobbet of barley 147 lbs., and a hobbet of oats 105 lbs.

Every encouragement is given to agriculture by the County Council. Exhibitions are awarded to enable young farmers to undergo a practical course of Agriculture at the University, as well as scholarships to attend a course of instruction in butter and cheese-making and general routine of dairy work at Llewenni Hall. Field experiments of a comprehensive kind are also carried out.

Under the Small Holdings Act the County Council has already purchased 690 acres at a cost of £26,416. The holdings are situated in different parts of the county and are divided into suitable allotments to meet the requirements of each district.

12. Industries and Manufactures.

The chief industries in the county are those of lead and coal mining, and agriculture; but as these have been considered at length in other chapters we need not refer to them here. The introduction of machinery and the conversion of arable into pasture land have caused many old industries to disappear. There were once on the banks of the Holywell stream iron-wire mills, a rolling mill, white and red lead works, three corn mills, three snuff mills (the snuff produced was in great demand), smelting works, a company for the manufacture of calcined "Black Jack," a pin mill, Parys mine copper works, large cotton works, brass-smelting works, as well as copper and brass works. At that time forty vessels were employed at Greenfield Quay to convey raw material and manufactured articles.

One of the largest factories in the county at the present time is the Hawarden Bridge Ironworks, where 5000 odd hands are engaged. The works comprise steel works, rolling-mills, sheet mills, foundries, and chemical works, as well as the galvanizing and corrugated-sheet factory. They cover over 60 acres of land and resemble a whole town of factories, each of these 30 tall smoke-stacks apparently indicating large separate and independent works, instead of being as they are, only a part of an immense whole.

The Glansztoff Works, situated at Flint, is one of the latest industries started in the county. About 60 acres of land have been acquired by the Company, and the works are planned to cover eleven acres. The industry, a good example of modern scientific methods, produces a kind of artificial silk, which is so skilfully manufactured that it is difficult even for an expert to distinguish it from real silk, and it takes a dye of any colour or shade.

Sandycroft Limited are large manufacturers of mining and electrical machinery of all kinds. The works established in 1855 at Rhyd-y-Mwyn, were removed twenty years later to the present site, where they occupy 17 acres on the banks of the Dee. Here are made stamps for the crushing of gold, silver, and other ores, winding and hauling engines, rock drills, etc., and these and other kinds of mining machinery are exported to all parts of the world, especially to the South African, Indian, Mexican, and South American mines.

The Dee Ironworks, Saltney, excel in the making of chains and anchors, which find ready markets all over the world. The Dee Shipbuilding Works stand on about three acres of ground at Queensferry, and here both steam and motor-driven vessels and steamers up to 250 feet long are constructed, a speciality being made of light-draught vessels for shallow waters (10 to 30 inches) and a unique type of stern-wheel steamers. These are largely exported to the Nile, and East and West African rivers and lakes, and over 200 vessels have been sent to the river Amazon and its tributaries. The Mostyn and Darwen Ironworks form a conspicuous object on the Dee estuary, especially

Stern-wheel Steamer

(*Built at Queensferry, being re-erected at Theresina, Brazil*)

at night when the large blast furnaces illumine the sky for miles around.

A large number of industries—chemical works, agricultural-implement works, paper mills, mercerising and bone-manure works and others—can only be alluded to here, but there are two manufactures of special interest.

In 1874 the Welsh Flannel Manufacturing Co., Ltd., was established on the St Winifred's stream. Welsh flannel has always been recognised as superior to any other flannel, owing to its wonderful softness. These are the largest textile works in Wales and they occupy three

Welsh Flannel Co.'s works

large mills with many acres of land adjoining. The machinery is driven by electricity and the looms afford work to 300 employés.

The other manufactory is the Hydraulic Cement Works at Afonwen Caerwys, which has been successfully carried on for years. Cement with the property of being able to set under water can be obtained from only three sources in the country: (1) by mixing white limestone with clay, (2) by burning the black limestone (Aberdo Stone), (3) by burning the cement stones (argillaceous limestone) of the Holywell shale. At the above works a

Cement Works, Caerwys

dark limestone from the neighbouring quarry is ground up, and mixed with rather less than a quarter of its weight of red clay (Boulder Clay). The mixture is then pressed under grindstones and passed through sieves and thoroughly mixed in a water-mill. It is then dried, cut up into blocks while soft, burnt, and then ground again, when it is ready for use.

13. Mines and Minerals.

Our county is rich in minerals, and its mines have been worked continuously from the time of the Roman occupation up to the present day.

The metal-yielding mines are confined to the Carboniferous Limestone and Chert. The following ores have been raised:—galena (sulphide of lead with silver), carbonate of lead, blende or "black jack" (sulphide of zinc), calamine (carbonate of zinc), haematite (oxide of iron) which contains nickel and cobalt in some parts; and copper pyrites and black oxide of manganese—the last two in small quantities only.

The Romans worked in Talar Goch and other mines in the county, but it is hardly probable that they ever attempted to reach the deep-seated veins which have yielded the bulk of the ore in modern times. The Roman coins and instruments unearthed in the old washes prove that the veins were most likely followed by means of trenches or levels with shallow shafts. The smelting was probably done near these open trenches, on the brow of a hill exposed to the prevailing winds. The Romans doubtless smelted lead extensively at Pentre Ffwdan (or according to Pennant, Pentre Ffwrndan, "the place of the fiery furnace") on the Dee, eleven miles from Chester.

The Limestone and Chert rest on the upper measure of a Silurian bed (Wenlock Shale) with an average dip of one in four to the north or north-east. They are covered by the Lower Coal Measures, and the most productive zone in north Flintshire lies immediately below these shales. The influence of the shales on the distribution of the ores seems to have been partly due to their having acted as water-tight blankets, checking the flow of underground water, which has taken an easterly direction ever since the rocks acquired their present dip eastward. But the precipitation also of the ores of lead and zinc was probably due directly to the action of sulphuretted hydrogen given off during the decomposition of pyrites in the shale, as the smallest trace of any salt of lead or zinc in solution would be arrested on coming into the presence of this gas. The ores of lead occur in the Chert beds, and in the upper and middle beds of the Carboniferous Limestone, while haematite has been found only in the lower beds of the limestone. The veins divide themselves into two distinct sets—the "veins" running east and west, and the "cross-courses" running north and south. The veins differ from the cross-courses in their contents as well as in their direction. They contain blende and galena, the latter with a proportion of silver ranging up to 14 ozs. or very occasionally up to 18 ozs. per ton. The cross-courses, on the other hand, contain no blende and very little silver, but frequently show copper pyrites. The cross-courses, which are often empty or in communication with large empty caverns, act as the principal underground water-channels and are considered to be of later date than the east and west veins. At a depth of 205 yards in these courses logs of timber with a girth of 1½ feet were discovered.

The fissures of veins so frequently found in the county were probably formed either by disturbance or by contraction. Their depth is unknown. Up to the middle of last century it was thought that the measures ceased to be productive at about 140 yards below the surface, but mining operations in the county have not yet revealed their limit in this respect.

The deposit of lead in the fissures varies from ¼ inch to three feet in width. A rib of two inches wide is considered to be remunerative, while one of four or six inches pays well. The vein varies in depth up to 100 yards. The lead is found in "lodes," which are nearly vertical, and "flats," which are horizontal, and vary in thickness like an ordinary seam of coal.

At different periods mines productive of vast wealth were tapped in several parts of the limestone and chert formations. One of the richest discovered was that of Pant Pwll Dwr, in Halkyn Mountain, which within 30 years enriched its proprietors by more than £1,000,000 sterling.

The Talar Goch mine, worked to a depth of 360 yards, was rich in lead and blende. The amount of galena produced in 40 years amounted to 57,752 tons, yielding 43,320 tons of metallic lead, with an average of 9½ ounces of silver to the ton; and the yield of blende during 30 years was 50,000 tons.

The Merllyn vein near Whitford, was worked in 1850 to a depth of 70 yards, but a quantity of ore reputed to be of the value of £150,000 was raised from a depth of about 20 yards only.

The mines richest in silver are the Holway mine, which produced 16 to 18 ozs. to the ton, and the Waen mines, Whitford, which produced 11 to 12 ozs. to the ton. The portion of the Chert escarpment known as Pen-yr-Henblas Halkyn has proved exceedingly rich in lead ore, and very profitable lead mines are now being worked at Halkyn and Rhosesmor.

Lead Mine, Halkyn

Blende is found in large quantities in different parts of the county. It is in metallic form and of a bluish-grey colour. Calamine is very plentiful in the county. Some 200 years ago, when the inhabitants were ignorant of its value, it was used for the mending of roads, but since then it has been worked most successfully in many places, though it is almost entirely confined to the eastern side of the county. It assumes various shapes and colours—green, yellow, red, and black. It often has a spongy appearance, and the richest looks like bees-wax.

The ore of iron lies in the lower beds of limestone only, never occurring in the same vein as the ore of lead and zinc. It sometimes contains traces of copper, and usually occupies chambers or irregularly widened-out spaces in north and south joints. The ore seems to have been formed in the replacement of the carbonate of lime of the limestone by the carbonate of iron which, by weathering and the escape of carbonic acid gas, has subsequently become converted into oxide. It is found at Bryn Sion, west of Ysceifiog, and near Caerwys. Haematite is believed to have been worked on Moel Hiraddug from a very early date; probably by the Romans. The ore occurs in nodules of dark, hard, and almost pure peroxide of iron in the spaces that run north and north-west. The Marian mine near Cwm is stated to have yielded over 28,924 tons in fifteen years.

Haematite with cobalt and nickel, which occurs in cavities or irregular chambers in the limestone, has been worked on Moel Hiraddug. But nickel and cobalt were not present in any quantity. Haematite with copper has been obtained in small quantities in Graig Fawr, and in the Diserth mine a deposit of iron pyrites was found.

Pyrolusite (the black oxide of manganese) occurs as an amorphous powder in the Pentre quarry, near Gronant and Moel Hiraddug, also in the north Hendre mine in a bed of "tuft" and yellow clay, but not in large quantities.

Up to the beginning of the seventeenth century

mining operations had been carried on very successfully by means of primitive appliances—namely the hand turn-tree, the horse and pulley, and the horse whincey. These came to a standstill owing to water difficulties. Then colonies of miners from Derbyshire and Cornwall settled in the county and introduced "day adits" and "levels" with very satisfactory results for a considerable period. Later, steam winding and steam pumping engines (known as "fire engines") were introduced, to be superseded eventually by the more powerful "Cornish pumping engine." At the end of the last century—the golden period of lead mining—there were a large number of these engines in the county. Mining has been much impeded in the county owing to the presence of water, and several profitable lead mines have been abandoned. To overcome this obstacle an Act of Parliament was obtained in 1875 for the purpose of enlarging and extending the work of the Halkyn Drainage Company. This level, which drained the Halkyn Mountain, brought in its train a period of great prosperity. In 1896 the Holywell-Halkyn Mining Company was formed to drive a tunnel from the lowest possible level on the bank of the river Dee through the rich mineral district towards Halkyn Mountain. The tunnel begins at the Dee Bank Wharf at Bagillt, nine feet below high-water mark. Its entrance is protected by two self-acting doors. For the first 8970 feet it is lined with bricks, for 150 feet with cast-iron tubing, and the remainder penetrates the solid rock. The tunnel is circular, eight feet in diameter, with a waterway six feet wide by one and half feet deep and covered by six inches of flooring timbers. It is equipped with a tram line, over which the minerals are drawn by ponies to the dressing rooms, and as the gradient is but 1 in 1000 from the entrance a pony can bring out 25 tons of ore a day. The Milwr Tunnel, started in 1897, is now nearly three miles in length and has struck several valuable lodes of lead and zinc ores. The driving of the tunnel costs on an average about £4 a foot. In 1913 an Act of Parliament was obtained by the Halkyn Drainage Company empowering them to continue the Milwr Tunnel into their area for a distance of about five miles, at a depth of 75 yards below the tunnel carried out under the Act of 1875.

14. Fisheries and Fishing Stations.

The fisheries of the British Islands form one of our most important industries, providing regular or occasional employment for nearly 100,000 men and boys. It is said that eight million pounds worth of fish is brought into the ports of England and Wales annually. The most productive of our fishing grounds is the North Sea, an area of more than 150,000 square miles. This large fishing-ground provides the bulk of the fish sold in our markets and more fish is brought every year into Grimsby and Hull, Lowestoft and Yarmouth than into all the other fishing ports of England put together.

The amount of fish caught along the coast of Flint is considerable, though of course with such a small coast-line it can hardly be compared with the important stations on the east coast, and the number of persons engaged in this industry in our county is not large. About a dozen

A Hoylake Trawler

fishing-boats anchor at Mostyn, but on account of the continual changing of the Dee channel most of the fishing fleet is stationed on the Cheshire side of the estuary, at Hoylake, Parkgate, and Heswall. The Hoylake fleet of trawlers is the most important in the estuary of the Dee. It consists of 39 first-class trawlers and 12 inshore fishing-boats. The first-class boats are yawl-rigged smacks of about 40 tons register and about 60 feet in length. They carry beam trawls of about 50 feet spread, and have steam capstans for hauling in the trawl and hoisting the sails. These never fish in the Dee estuary but frequent deep water off the coast of North Wales. The second-class boats stationed at Heswall and Parkgate do most of the fishing along the coast. They are half-decked boats of about 10 tons register and carry a trawl of some 25-feet spread, which is used for taking flounders and shrimps in shallow water.

The most important food-fishes that frequent the Dee estuary are the plaice, sole, flounder, dab, halibut, turbot, whiting, haddock, cod, and sprat. Herrings and mackerel are also occasionally caught in great numbers. The coast abounds with young plaice and often with the young of sprats, herring, and other food-fish. Unfortunately the shrimp-trawling, largely carried on throughout the year in the Dee, is responsible for the death of countless thousands of them, for even if thrown overboard immediately they often fail to survive, being crushed and killed by the weight of other fish caught with them.

The sea fisheries are regulated by the Dee Sea-Fisheries Committee, which is supported by the various County Councils, whilst the fresh-water fisheries are managed by the Dee Board of Conservators. The committee have complete control over the fishery and they issue regulations directing, among other things, the kind of nets that may be used and the size of their meshes for various fish, as well as regulating the size of fish which may be taken.

The Dee is famous for its salmon and a large number are caught in it annually. The portion of the river above

Ysceifiog Lake

the Bangor bridge in Maelor is considered one of the best spawning grounds. The fish are captured by strike-nets floated from a boat, by coracle nets, and by rod and line. The best fish are taken in the "straight cut" below Chester.

The Elwy and Clwyd are also excellent for trout and salmon, over 1000 of the latter fish having been netted in 1910, and about 300 landed with the rod; the average weight of the fish was 5 lbs. The Alun and the Wheeler abound with trout, and Llyn Helyg with pike, perch, and roach.

In 1904 trout hatcheries were formed by the Earl of

Landing a Rainbow Trout, Lake Ysceifiog

Denbigh at Ysceifiog, Downing, and Gwibnant. That at Ysceifiog consists of 40 ponds, and a lake about 20 acres in extent was constructed for angling purposes and storing the fish reared in the hatcheries.

Sturgeon are not infrequently caught in the Dee and some curious formalities have to be gone through by the fishermen before they may sell the "royal fish."

15. Shipping and Trade.

The shipping industry in the estuary of the Dee is not what it used to be in former years, and the Board of Trade returns show an annual decline in its trade.

History tells us that when Edward I left Chester to conquer Wales, his fleet sailed down the estuary to guard him and to carry provisions for his men. Chester was then an important port, the Clwyd was navigable up to Rhuddlan, and all the ports along the coast were accessible even at low water. Rhuddlan and Flint are constantly mentioned in the reign of Edward II among the important harbours of the island. When the Dee estuary, with Chester as its chief port, was at the zenith of its prosperity, the estuary of the Mersey was yet in its infancy in the commercial world. Where once the bulrush flourished stand the immense harbours and docks of Liverpool and Birkenhead. While Chester is now visited by only a few coasting vessels and floats, Liverpool is the cotton port of the British Isles.

This change is due largely to the continual silting-up of the Dee estuary, a process which has been allowed to proceed uncontrolled for centuries. This tendency to silt up occurs also in the Mersey, but constant and careful dredging has kept this river navigable. The sea gains considerably on the land towards Rhyl and Prestatyn by eroding the soil. Much of this soil is carried up the Dee by the strong tides driven by prevalent north-westerly winds. The flow of the tide is rapid and the slow ebb precipitates in the river bed all the suspended matter, chiefly sand and clay.

From the Board of Trade returns for 1910 we learn that nearly 60,000 tons of iron and manganese ore and iron pyrites were brought to the various ports of the river, namely, Rhyl, Mostyn, Connah's Quay, and Chester, to

Bagillt Harbour

(*Now of little importance through the silting-up of the river*)

the value of over £80,000, together with 2300 tons of wood and timber, valued at £6000. The principal exports consist of chemicals and chemical preparations.

The number of foreign trading vessels that entered our ports during the same year was thirty-nine, of which twenty-nine were steam vessels. Again, 1410 British and foreign sailing and steam vessels (their repeated calls included) entered with cargo and ballast, coast-wise. Their cargoes and ballast amounted to 103,892 tons. The number and tonnage of vessels arriving at the various stages in the river for the year 1910 showed a decrease compared with the returns of former years. The income from duties in Connah's Quay in 1909 was £528, in 1908 £604. Ten years ago the tonnage of the river was 78,000, while now it is hardly half that figure.

Traffic is greatly impeded owing to the difficulties experienced in navigating the river, and trade consequently suffers severely. Of the natural harbours along the estuary of the Dee, Connah's Quay is the most important. Ships call also at Greenfield, Mostyn, Point of Ayr, and Foryd.

In 1832 we find that over 800 ships entered the port of Flint, including six vessels from America. Before the construction of the stone bridge at Rhuddlan in the sixteenth century, the river Clwyd was navigable to the castle walls, and as late as seventy years ago vessels of fifty tons could sail at high water to the bridge. Much traffic was carried on at the time, especially in agricultural produce such as barley and wheat. But when Foryd railway bridge was constructed the shipping at Rhuddlan disappeared.

16. The History of Flintshire—(*a*) From the Coming of the Romans to the Norman Conquest.

Before the birth of Christ the Roman Empire had arisen in the path of the nations migrating westwards. The Romans were not merely settlers like the waves of nations that had passed over Europe, they were conquerers and rulers. They established a vast empire of settled government, and the nations, for a time at any rate, ceased to wander westwards.

The Romans fixed upon the site of Chester as their principal military centre in the west of Britain, and it was the home of one of their legions for over 400 years. The land beyond the estuary of the Dee was in sight of the walls of Chester; and it seemed as easy of administration as, apparently, it had been of conquest, an event which took place during Suetonius Paulinus's famous march to reduce Mona (Anglesey), so vividly described by the Roman historian, Tacitus.

The Romans constructed a road across the county from Deva (Chester) to Segontium (Carnarvon). They built a villa at Ffrith, in the parish of Llanfynydd, and possibly made a road crossing it to Caersws in the county of Montgomery. The minerals of the county attracted their attention, and they carried their lead to a smelting centre at Pentre, near Flint.

While outlying provinces like Britain had become prosperous and had not yet been ground down by the increasing severity of the Roman system of taxation, the Roman Empire began to decline from internal decay. In the fifth century Rome fell before the barbarians, and the hardy hordes of the north and east again began to press on westward.

Britain was invaded by Teutonic tribes from modern Denmark and Germany. They conquered the east and the north of Britain, while the Angles advanced from the Humber, and the Saxons from the Thames and the south coast. It was almost a race between them for Chester. The Saxons penetrated along the valley of the Severn, but being checked in the decisive battle of Faddiley, they were forced to retire. This gave the Angles of the north their opportunity. Under their great king Ethelfrith they advanced towards Chester, and about 615 the critical battle was fought described by the English historian Bede, who probably got his account from the sons of those who were present. Ethelfrith after some indecision was victorious. Chester fell and ceased to be of any importance in the history of the country for nearly 400 years, the district now called Flintshire being no longer protected by it or ruled from it. It became the battle-ground between two powers, which were gradually developing, the kingdom of Gwynedd and the kingdom of Mercia.

In 796, the year of the death of Offa, a battle was fought at Rhuddlan between the Mercians and the Welsh. Fifteen years later Cenwulf, Offa's successor, died at Basingwerk when defending his frontier. The fall of Chester in 615 had left the Dee exposed to the ravages of the nations that attacked Britain on the west, and during the struggle between the Welsh and the Mercians, the Danes stepped in and acted as allies first for the one side and then for the other. When Mercia and Wessex fought for the overlordship of Britain, the Welsh were at first glad of an opportunity of attacking their old enemies; and their success culminated in the victories of Gruffydd ap Llywelyn. This king had his house in Rhuddlan, where he kept a fleet to guard the coast and an army to cross the dyke. He had also a residence at Bistre to watch over his subjects in the Alun Valley. He regained the district from the Dee to the Clwyd and forced the English settlers to leave their homes which they had called Preston (Prestatyn), Merton (Mertyn) and Whitford. He advanced against the Mercians and defeated them at the battle of Rhyd-y-Groes in the valley of the Severn about the year 1039. His influence grew and soon he found himself facing the rising power of Wessex. Gruffydd entered into an alliance with Mercia, and thus healed the old feud which had divided modern Flintshire for so long. But the alliance brought a new enemy into the district. Harold the son of Godwin, now the ruler of England as well as the Earl of Wessex, made a forced march to Rhuddlan. Gruffydd had time to escape in one of his ships, and Harold had to retire. Gruffydd's plan was to avoid a pitched battle with Harold's well-appointed army, to allow him to starve in the country which he had harassed, and to attack him during his retreat. The people of the country, however, had not the patience to carry out this policy, and traitors killed

Rhuddlan Castle

the great king. The district became subject to England, and Harold placed it under the rule of Gruffydd's half-brothers, Bleddyn and Rhiwallon.

Wessex did not remain long in the possession of the overlordship. In January, 1070, William the Conqueror appeared at Chester, and the town was given eventually to an able and powerful baron, Hugh of Avranches—"the Wolf." Hugh entrusted the conquest of the country westwards, especially modern Flintshire, to Robert of Malpas and Robert of Rhuddlan. The conquest of the English and the Welsh was carried out with the greatest ferocity, and the pages of Ordericus Vitalis, himself of Norman blood, describe Robert of Rhuddlan as a veritable monster. The whole area occupying the modern county was soon brought into subjection, and Norman castles rose to mark the conquest at Hawarden and Rhuddlan.

(*b*) Tegeingl in Domesday Book.

After William had conquered England in 1066, he appointed commissioners to make a survey of his newly acquired territory. The records of this inquiry are known as Domesday Book. As the district now called Flintshire is fully described therein it was evidently in the possession of the Normans in 1086, and we are enabled to form some idea of its condition at that period.

The chief division was called the Hundred of Atiscross, which was practically equivalent to Englefield. It contained about eighteen manors. These manors comprised so many "hides" of land, and contained large forests : the one at Soughton was estimated as ten leagues by three leagues, but the other twenty-four "woods" mentioned are on an average one square league in area.

Roelent (Rhuddlan) had its mint, its church, and its fisheries. About twenty-two "berwicks" belonged to the manor, some of which were situated in Ysceifiog, Cilcain, and Diserth. In these berwicks stood the mud cabins and thatched homesteads of the people, and around them grew the crops of bere, a coarse grain resembling barley, which was the staple food of the inhabitants, and from which the berwick took its name.

As was the case in England the tenants in chief were mostly Normans ; the Welsh and English peasants became sub-tenants. The various classes of society recorded in this district in Domesday Book are as follows—tenants-in-chief, 12 (these were freemen) ; bodars, 87 ; villeins, 74 ; burgesses, 18 (these lived in Rhuddlan) ; foreigners, 4 ; priests, 4 ; neatherds, 2 ; servant, 1. No reference is made to women and children.

The bodar was of the most numerous class : he lived in a small cabin made of mud and wattle, and held a small portion of land on condition that he supplied poultry, eggs, and other articles of food for the table of his lord. The villeins were labourers. Their duty was to carry manure, to hedge and ditch the demesne of their lord, and to do other humble work. They could acquire no property either in goods or land. The burgesses had special advantages granted them and facilities for trading. All these resided at Rhuddlan, "a new burg," and they had the privileges of the burgesses of Hereford, namely, "throughout the year they shall give nothing but twelve pence for any forfeiture save homicide, theft, and premeditated heinfare."

The serfs seem to have been of lower rank than the villeins and were reduced almost to a state of bondage. They might hold land but only at the will of their lord. Although only four priests are mentioned there were churches at Hawarden, Rhuddlan, Whitford, Diserth, Brynford, Gwespyr, Gwaenysgor (in ruins) and Prestatyn. The list of animals given as belonging to the district is very meagre. Eighteen oxen fit for the plough are mentioned. There was an eyry of hawks at Diserth and at Bistre, and four at Soughton, and there were fisheries at Whitford and Rhuddlan.

(*c*) From Gruffydd ap Cynan to the Statute of Rhuddlan.

Gruffydd ap Cynan, who was by birth half Welsh and half Irish, sailed up the Clwyd in 1075 to ask Robert of Rhuddlan for help to fight Trahaearn; but after securing Gwynedd he returned to attack Rhuddlan Castle and to kill many of its defenders. In 1088 he entered the Conway river with three ships to harass the district and to capture a great booty of men and cattle. Robert of Rhuddlan on being informed of this, accompanied by a single soldier, Osborn de Orgar, and without any defensive armour except his shield, hurried in pursuit. Gruffydd's men attacked him and cut off his head, and fastening it to the mast, sailed away in triumph. Thus was the first lord of the district disposed of. Six years later William Rufus invaded Gwynedd but had to seek an inglorious retreat.

After the death of Gruffydd ap Cynan, "the King, sovereign, prince, and defender, and pacifier of all Wales," his son Owain Gwynedd succeeded him, and vigorously pursued his father's policy. In 1146 his army appeared in this county with the object of taking Mold Castle, one of the strongest forts of the barons at this period. It fell, and four years later we find Owain Gwynedd marching from Rhuddlan to defeat Ranulf, Earl of Chester, who was supported by Madog, Prince of Powys, at Cynsyllt, near Flint.

In 1157 Henry II made a very determined effort to check Owain Gwynedd's triumphant career. The King had no opposition elsewhere and some of the Welsh princes who were jealous of Owain united their forces with those of Henry. Chester was made the military base, and the fleet was ordered to sail from Pembroke to the estuary of the Dee. While Henry was forming his camp on Saltney marsh, Owain Gwynedd posted his army at Basingwerk or "Dinas Basing," ready to check the King's army on its way to Rhuddlan, and sent his two sons, Dafydd and Cynan, with a small body of men, to lie in ambush in the great forest close to Ewloe Castle. Henry's main army travelled along the coast; but he himself with a body of his men took the forest route, and were successfully ambushed. Henry escaped, but Eustace Fitz-John, Constable of Chester, and other prominent barons were slain, and it was with difficulty that the survivors found their way to the shore to join the English army. Owain Gwynedd, satisfied with the result of his strategy, retreated to St Asaph. Henry II proceeded to Rhuddlan and eventually peace was established between him and Owain Gwynedd. Englefield once more fell into the hands of the English, Rhuddlan was entrusted to Hugh Beauchamp, and Basingwerk was fortified. The peace, however, was of short duration. In 1163, Dafydd, the son of Owain Gwynedd, entered Englefield and attacked Rhuddlan and Basingwerk castles ; and Henry II had to lead a hurried expedition to their relief. After the defeat of the English at the battle of Crogen in 1165 Owain Gwynedd made a determined effort to regain what is now termed Flintshire. He first made a successful attack upon the troops engaged in repairing Basingwerk, attempting later the more formidable task of recapturing Rhuddlan. This he succeeded in taking after a three months' siege and with the help of the men of Deheubarth. One of Owain's last acts was the destruction of Rhuddlan and Prestatyn castles.

In the beginning of 1199 Llywelyn the Great captured Mold Castle, the home of Robert Montalt, and a very important outpost of Hawarden and Chester, and nine years later his troops ravaged the estuary of the Dee and attacked Rhuddlan Castle, to the dismay of Ralph of Chester. Llywelyn had married King John's daughter, Joan, but his action aroused the anger of his father-in-law, and in 1211 Llywelyn with his men retreated to the fastnesses of Snowdon, King John meanwhile marching through Englefield to Bangor in pursuit. Through the intercession of his wife Joan a peace was made, and Englefield was acknowledged once more as English territory, but in the following year Llywelyn drove the royal stewards from Deganwy and Rhuddlan, and again secured the land between the Clwyd and the Dee. Roger of Montalt now took possession of Mold Castle, and another

Remains of Diserth Castle

castle was built in the place of the one at Rhuddlan, on the rock above Diserth.

Henry III invaded Tegeingl in 1241 against Dafydd ap Llywelyn, who had captured Mold Castle and was attacking the new castle at Diserth, at this time one of the strongest forts in the district. It was an exceptionally dry summer, and the royal army could move easily across the great marsh of Rhuddlan. Dafydd surrendered without fighting, and at Gwern Eigron on the Elwy, two miles south of Rhuddlan, he signed the humiliating agreement called the Treaty of Alnet, and Englefield was transferred to the English King. Henry III granted Chester, as well as the district between the Dee and the Clwyd, to his son Edward, who was later to figure so prominently in Welsh history. Thus in 1254 the region now called Flintshire fell entirely under the jurisdiction of Chester, and it appeared probable that it would be finally attached to the County Palatine and not formed into a Welsh shire.

Henry III's reign was disturbed by the Barons' War. Llywelyn ap Gruffydd, last Prince of Wales, joined Mortimer's cause and was promised Hawarden Castle as a reward, although it was not yet taken from the hand of Robert of Mold. After the defeat of the King Englefield was added to Mortimer's estates. When Simon fell at Evesham, Llywelyn was carrying fire and sword towards the gates of Chester. In 1263 we find Englefield in his possession, Diserth Castle destroyed, and Mold added to his conquests. In 1265 he besieged Hawarden Castle, which he took and dismantled, and Robert of Mold, one of the Lords Marcher, was among his prisoners. So powerful had Llywelyn become that the King in the Treaty of Shrewsbury acknowledged him Lord of the Four Cantrevs, which included the territory between the Dee and the Clwyd.

In order to defend his newly-gained possession, Llywelyn built a castle at the extreme end of Englefield, in the corner of the wood of Ewloe. So confident was he in his power that he allowed Robert of Mold his liberty and permitted him to repossess his land at Hawarden, on condition that he would not rebuild the castle for thirty years. Evidently Ewloe Castle was well garrisoned before he could venture to grant freedom to so formidable an enemy. Llywelyn retained Mold till 1269, so that the whole of modern Flintshire, except Robert of Mold's land at Hawarden, was Welsh at the death of Henry in 1272.

When Edward I became King he began a vigorous campaign in 1277 against Llywelyn, who had refused to pay homage to him. Edward started, like all other invaders, from Chester, cutting his way through the forest and making a road ready for his retreat. He moved cautiously through the district which had been the scene of the disaster of Henry II, and fixed his head-quarters at Flint on July 26th, where he met Dafydd, Llywelyn's brother, with whom he made a treaty, and promised him a proportion of his brother's land for his support—Dyffryn Clwyd and Rhufoniog as well as the castle and land of Hope.

Edward fortified Flint and Basingwerk, and on August 21st reached Rhuddlan accompanied by the queen. The English army that assembled along the Clwyd numbered over 15,000 men. Dafydd brought only 200 infantry and 160 crossbow-men, and nothing remained for Llywelyn but to negotiate for peace. On the banks of the Conway a treaty was accordingly signed; Llywelyn was compelled to pay homage to Edward I and to cede the Four Cantrevs.

To check further attacks on Englefield, Flint Castle was built, and Rhuddlan Castle strongly fortified, and the district again came under the jurisdiction of Chester. All this was done to a great extent under the personal supervision of the King.

The Treaty of Conway was soon nullified, for Reginald

Flint Castle

de Grey, in direct violation of it, introduced new customs to Englefield, and the people moreover suffered from the corruption and oppression of the justices and bailiffs. In 1281 the men of Flint brought their grievances before the King, and next year Dafydd openly deserted the King and took sides with his brother Llywelyn. On the eve of Palm Sunday, 1282, he made a successful attack on Hawarden Castle, and by the end of March the cantrevs were up in arms espousing Llywelyn's cause. This brought Edward I to Englefield to undertake the second Welsh War.

Edward's base of operation was now at Rhuddlan, not Chester, with a garrison at Flint. Dafydd was in possession of land running parallel with Englefield on the south and on the east side of the Clwydian Hills as well as Hope, which left his antagonist in a dilemma, for he could not move west while there was so formidable an enemy on his flank. He accordingly summoned his barons to his aid to engage Dafydd, and ultimately the Justice of Chester, Reginald de Grey, captured Ruthin and Dyffryn Clwyd, the Earl Warren became the owner of Bromfield and Yale, and Henry de Lacy conquered Denbigh. The success of these operations, it should be remarked, accounts to a very great extent for the irregular shape of our county, and probably had it not been for Edward I's trouble with Dafydd, Flintshire would not be a Welsh county but would have remained as before a part of the County Palatine of Chester. The result was that Edward lost three of the Four Cantrevs, and allowed three Lords Marchers to settle on land that he once owned as Lord of Chester. These remained independent rulers till the time of Henry VIII, and were a constant source of trouble to the English Kings.

Later Edward I, supported by the Earl of Lincoln, moved from Rhuddlan, and Reginald de Grey from Hope, and their united forces marched towards Gwynedd, the fleet of forty ships meanwhile leaving the mouth of the

Clwyd for Penmaen Mawr. The Welsh, however, made a vigorous descent from the slopes of Snowdon and inflicted a crushing defeat on their antagonists at Conway. But success was not destined to attend them long. Llywelyn was shortly afterwards killed on his way to rally South Wales, and Dafydd, whom he had left in command of his forces, did not long survive him. He held on in Eryri for a time, and afterwards moved to Castell-y-Bere near Cader Idris, and later to Snowdonia, where he was captured. This brought the war to an end. Early in September he was escorted through Rhuddlan by sixty archers, to suffer a traitor's death at Shrewsbury.

(*d*) From the Formation of the County in 1284 to the Great Civil War.

In 1284 the "Statute of Wales" was issued at Rhuddlan. This, usually known as the Statute of Rhuddlan, settled the government of the Principality until the Act of Union of 1536. By this Statute Englefield, which had been up to 1282 connected with Chester, was for the first time called Flintshire, and added to the Principality with the other five counties of Anglesey, Merioneth, Cardigan, Carmarthen, and Carnarvon. Thenceforward it had its own sheriff at Flint, who controlled the cantrev of Englefield together with the land of Hope, the land of Maelor Saesneg, and the land adjoining the castle and town of Rhuddlan as far as the town of Chester.

The Flintshire of Edward I was much smaller, and its boundaries more irregular than the Flintshire of to-day. Mold and Moldsdale were in the possession of Robert of Mold, and formed one of the Lordship Marchers, and the east part of Hawarden parish belonged to Chester. St Asaph, although no mention of it is made, was probably included in Flintshire. If a border-line were drawn from St Asaph through Ysceifiog and Buckley to Shotton it would enclose the Flintshire of 1284.

In 1322 Sir Gruffydd Llwyd of Tre Garnedd Fawr in Anglesey, who had previously been in the service of Edward I, and who brought him the news to Rhuddlan of the birth of his son Edward at Carnarvon, fought against the English and was captured, imprisoned, and put to death at Rhuddlan Castle.

In 1385 Richard II undertook an expedition to Ireland. The Duke of Lancaster took advantage of his absence, and returned from exile to England in 1399. Crossing over from Ireland Richard landed at Conway in September. As he made his way in company with his friends through a defile in the mountainous district near Abergele he was surrounded by Percy and his forces and taken captive. After a halt at Rhuddlan the royal prisoner was conveyed to Flint Castle[1].

In the rebellion of the Welsh under Owain Glyndwr Flintshire was divided. Glyndwr had married a Flintshire lady, Margaret, the third daughter of John Hanmer of Maelor. Tradition relates that Owain Glyndwr was

[1] Shakespeare in *Richard II*, Act III, Scene III, refers to this incident.

in the retinue of Richard II as he entered Flint Castle, and that he set fire to the cathedral of St Asaph to avenge himself on the bishop, John Trevor, for pronouncing the sentence of deposition on the King. In 1403 Owain's army was posted on Moel-y-Gaer, but his

Richard II and Henry of Lancaster at Flint Castle

enemies from the town of Flint captured and beheaded his chief, Howel Gwynedd. Flint Castle remained in the hands of Henry IV. Owain's most faithful allies lived about Hope and in Maelor, and to these and other of his supporters Henry granted pardon. Among them was one Owain ap Alltud, who had led the revolt in Flintshire for three years.

At the end of the Wars of the Roses, Flint took an active part in placing Henry VII on the throne. We are told that Henry spent some time at Mostyn Hall, and that he narrowly escaped being captured by the Yorkists by jumping through a window. Some 1600 Flintshire miners and colliers, the main force led by Richard ap Howel, the ancestor of the Mostyns, and another contingent under the Salesburys, marched to Bosworth to fight for Henry. Sir William Stanley and his men at a critical moment deserted Richard III, and took their stand with Henry Tudor.

When the Civil War broke out between Charles I and the Parliament, Flintshire, like the majority of the Welsh counties, declared for the King. Charles I, when he was at York in 1642, received an address from Flint full of strong expressions of loyalty. The county also paid £730 as an assessment for the support of the royal army. Sir Roger Mostyn was one of the first to take up arms for the King. He raised an army of 1500 strong, which he maintained at his own expense. In 1643 Cromwell's army of 2000 men, under Myddleton and Brereton, took Hawarden Castle and proceeded to Flint, which after a long siege was yielded upon honourable terms.

At the beginning of the winter of 1643 about 2000 Irish Royalists landed at Mostyn bedraggled and in a sad plight. The Flintshire castles were alternately in the hands of the Parliamentarians and the King. In 1645, after a forced march, Brereton's army posted itself at Hawarden and endeavoured to besiege the castle, which contained the King's ammunition. After witnessing from the walls of Chester the battle of Rowton Heath and the defeat of his men, Charles I called with a sore heart at Hawarden on his way to Denbigh. A month later, on October 30, 1645, the Parliamentary forces, under Colonel Michael Jones, reached Mold. They consisted of 1400 horsemen and 1000 footmen, "the cream of all those parts of the kingdom." After their march through the county they defeated the Royalists at Denbigh. During March, 1646, Flint Castle was once more in the hands of the Royalists, while Rhuddlan was one of the last four castles to hold out for the King.

In the second civil war Flintshire with the rest of Wales decided for the old Royalists and Parliament against Cromwell and the Independents, and Rhuddlan was rendered untenable. Sir Roger Mostyn was taken prisoner on May 10th, 1650, at Conway, but was soon liberated upon his promise "to be at his own house at Mostyn, and to engage in nothing prejudicial to the present government." His support of the King meant sacrificing a good deal; to this end he spent about £60,000, an immense sum of money in those days. His house also was stripped of its valuable possessions and he was obliged to live for several years at Plas Ucha, a small farm-house in the precincts of his old mansion.

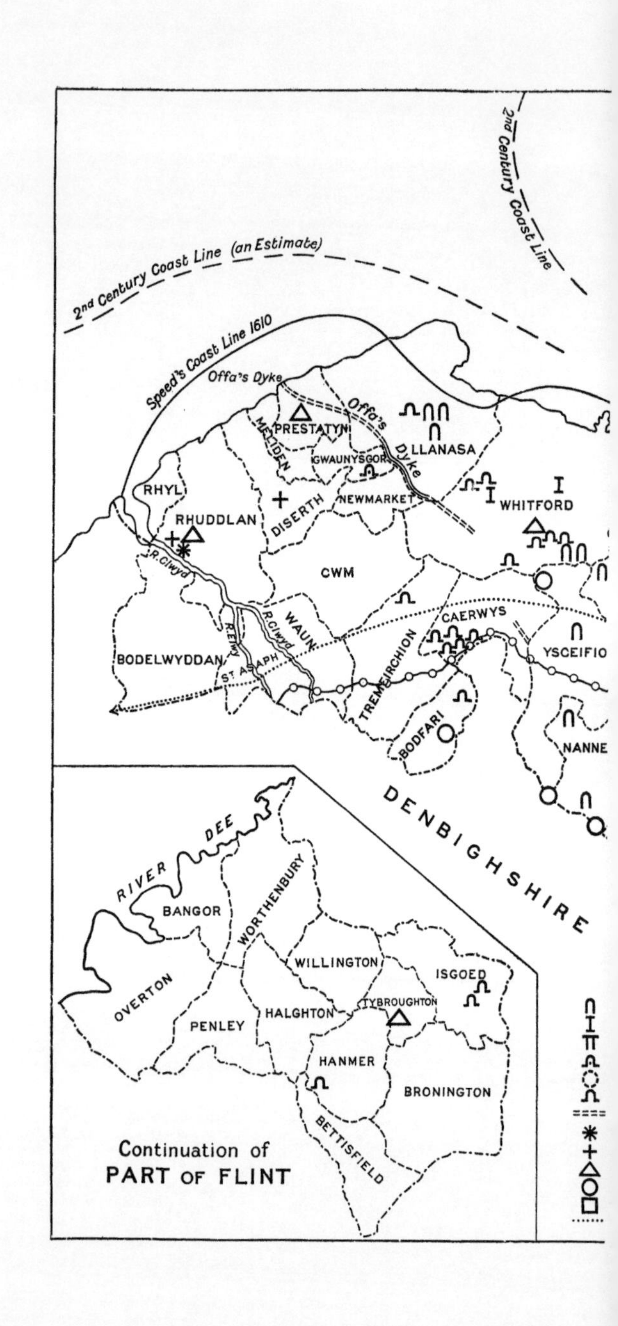
2nd Century Coast Line
2nd Century Coast Line (an Estimate)
Speed's Coast Line 1610
Offa's Dyke
Offa's Dyke
PRESTATYN
MELIDEN
LLANASA
GWAUNYSGOR
RHYL
DISERTH
NEWMARKET
WHITFORD
RHUDDLAN
R. Clwyd
CWM
WAUN
R. Elwy
R. Clwyd
CAERWYS
BODELWYDDAN
ST ASAPH
TREMEIRCHION
YSCEIFIO
BODFARI
NANNE
DENBIGHSHIRE
RIVER DEE
BANGOR
WORTHENBURY
WILLINGTON
ISGOED
OVERTON
TYBROUGHTON
PENLEY
HALGHTON
HANMER
BRONINGTON
BETTISFIELD
Continuation of
PART OF FLINT

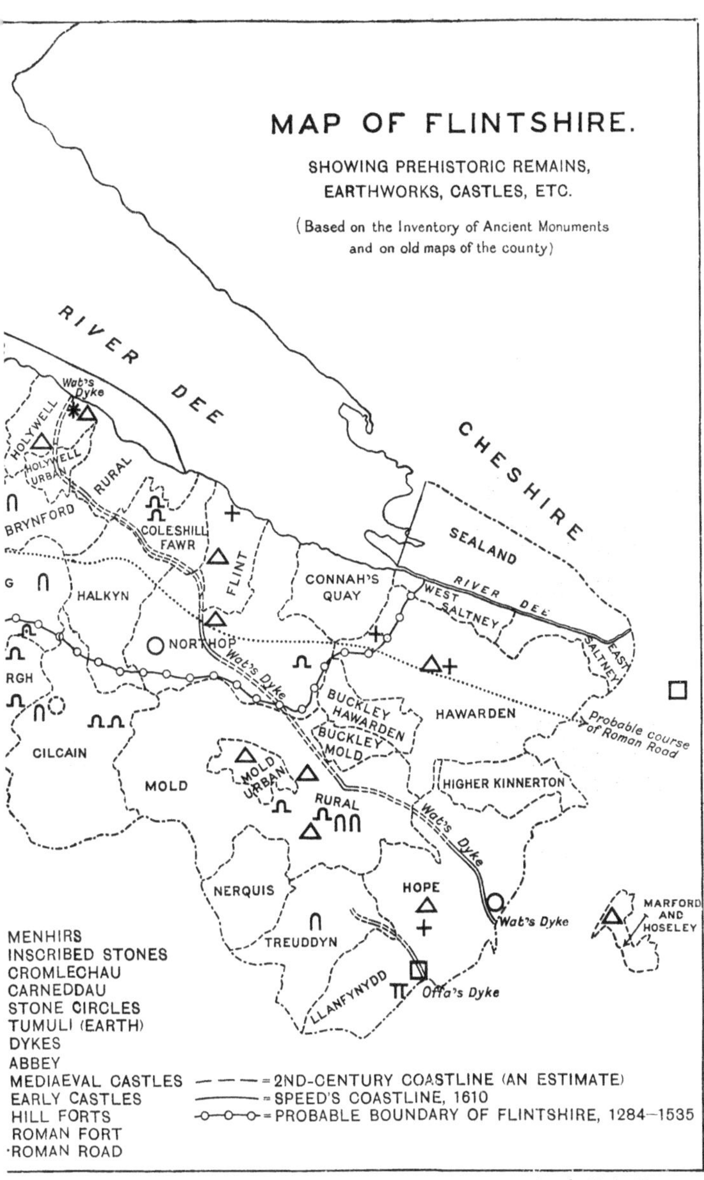

Camb. Univ. Press

17. Antiquities.

Of Palaeolithic man there are no traces in our county, but of man of the Neolithic or New Stone Age there is

Neolithic Stone Celt

(*Discovered at Gop, Newmarket*)

abundant evidence. The Iberian has left behind him his bones, his graves, and his stone implements as relics. From his bones we can tell that his people were short of stature with long skulls, prominent features, and narrow foreheads. His tumuli or graves are very numerous on the table-land between Newmarket and Caerwys; indeed, to find a district so rich in remains of this period as the plateau of Flintshire would be a difficult task.

The tumuli have been explored from time to time, and

Newmarket Gop

the remains in them have been generally found to consist of calcined bones and pieces of rude pottery. The Gop Cave in Newmarket clearly shows that it was the abode of man in the Neolithic period. In one of its chambers fifteen skeletons were found closely packed together in a crouching position. In this cave were also found a flint flake of rare type and a very perfect stone axe of peculiarly hard stone, nearly nine inches in length and four inches broad, beautifully chipped but not polished. Ffynnon Beuno Cave with its stalagmite floor and Cae Gwyn Cave, both near Tremeirchion, also had a great variety of bones and flint implements, all belonging to the Neolithic Age.

The Bronze Age brings us nearer to historic times, but it extends over a period of 1500 years. As relics of this period three objects of great value have been found: a gold torque, the Caergwrle cup, and a gold horse

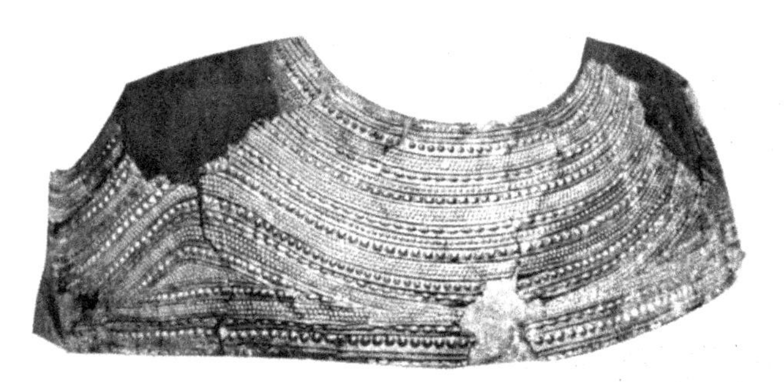

Gold Breastplate found at Pentre Hill

breastplate. These objects prove to us that the district now called Flintshire could then boast of wealth and of skilled craftsmanship of no mean order. By reason of these discoveries, of their intrinsic value and their unique character, Flintshire figures prominently in the history of Celtic Art. It is not known precisely where the gold torque, which weighs 24 ounces, was found, but probably either at Bryn Sion near Caerwys during the clearing of an old tumulus, or in a limestone quarry, near

Holywell. The Marquis of Westminster paid 200 guineas for it in 1815. The Caergwrle cup, now in the British Museum, was found in 1820 by a workman engaged in draining a field near Caergwrle Castle. It is an oval cup of black oak about 9 inches in height, the exterior being neatly overlaid with beautifully chased gold in various devices.

The gold horse's breastplate, which is now in the British Museum, was discovered in 1833 on Pentre Hill, sometimes called Bryn-yr-Ellyllon (the Elves HIll), about a quarter of a mile from Mold. Its width at the centre measures 8½ in. and its length about 3 ft. 7 in., but it is not intact at either end. It was the breastplate or peytrel of a pony apparently not more than 12 hands high. The thin gold plate was elaborately embossed and worked by skilful hands. Gold at that time could not have been scarce in our area, otherwise the warrior buried at Mold would not have decorated his horse with the valuable metal.

During this period the Meini Gwyr or Bowing Stones and Meini Hirion (menhirs) were erected. There is only one stone circle in our county, and it is situated in Pen Bedw Park near Nannerch on the Mold road. It is 29 yards in diameter and contains five stones varying in height from one to five feet, forming a segment of a circle. The other stones which completed the circle have been carried away.

Although the Roman XXth Legion was stationed at Deva (Chester), on the borders of the county, Roman remains are not numerous. But it is certain that a

Roman road traversed the district from Deva to Segontium (Carnarvon), though its path has not yet been fully traced. The Antonine Itinerary places a Roman station between Conway and Chester, and according to its mileage this camp should be on the western side of Flintshire. Whether the Romans had a camp at Caergwrle, Caerwys, Bodfari, and Pont Ruffydd, where Roman bricks and tiles have been found, is still a matter of conjecture.

Recent discoveries have fixed Ffrith in the parish of Llanfynydd as a Roman villa or camp ; walls, fragments of hypocaust tiles, portions of Samian ware, bits of urns and several objects of Roman design, as well as a Roman altar have been found there. But the rich minerals tempted the Romans, and we are safe in stating that they had smelting places at Croes Ati, Pentre, from the pieces of lead and tools they left behind.

From the withdrawal of the Romans to the coming of the Normans, A.D. 410–1066, there is but little to record. It was the period of destruction. There is nothing left of the extensive monastery of Bangor-is-y-Coed, and of the old church built at St Asaph in the fifth century. Of the churches mentioned in Domesday Book the names only remain. There are, however, a few sculptured crosses of special interest in the county.

Diserth cross, Croes Einion, once stood in the churchyard, but it has now been placed on a new stone pedestal within the church. It shows Celtic interlacing work and is about 6 ft. 6 ins. in height. When Llywelyn the Great was attacking Diserth Castle in 1261 Einion Rhirid

Maen Chwyfan

Flaidd was slain in the struggle and tradition says that a sepulchral cross was erected to his memory.

Maen Chwyfan is a fine wheel cross of the tenth or the eleventh century and stands in a field in Whitford parish. It is about 12 ft. in height, and is the tallest wheel cross in Great Britain. It is ornamented on all four sides. Chwyfan was a Welsh saint to whom Diserth church was dedicated. There are churchyard crosses at Cilcain, Diserth, Flint, Newmarket, Overton, Tremeirchion, and Ysceifiog; and wayside crosses once stood at Bangor, Criccin (Rhuddlan), Croes Wian (Caerwys) and Overton. The name "Maes y Groes" at Bangor, Bodfari, Cilcain, and Holywell suggests that a cross was erected on the fields so named in each of these places.

The cromlechs also belong to this period, but there is no cromlech in the county now, as the one in Hope parish has been lost. This was known as the burial place of Gwrle Gawr, who was supposed to have been connected with Caergwrle Castle.

Several urns have been found from time to time when excavating the numerous caves and tumuli. Two sepulchral urns were dug from a tumulus at Rhydwen, a farm in the parish of Caerwys. These are whole and in good condition and serve as an example of the pottery made and used in the county in the late Bronze Age.

18. Architecture—(*a*) Ecclesiastical.

The Pre-Norman or, as it is usually though with no great certainty termed, Saxon building in England was the work of early craftsmen with an imperfect knowledge of stone construction, who commonly used rough rubble walls, no buttresses, small semi-circular or triangular arches, and square towers with what is termed "long and short work" at the quoins or corners. It survives almost solely in portions of small churches.

The Norman conquest started a widespread building of massive churches and castles in the continental Romanesque style, which we term Norman. This period of architecture is distinguished by walls of great thickness, semi-circular vaults, round-headed doors and windows, and massive square towers.

From 1150 to 1200 the buildings became lighter, the arches pointed, and the science of vaulting was perfected by which the weight was borne by piers and buttresses. This method of building, the "Gothic," is the result of the endeavour to cover the widest and loftiest areas with the greatest economy of stone. The first English Gothic, "Early English" architecture, from about 1180 to 1250, is characterised by slender piers (commonly of marble), lofty-pointed vaults, and long, narrow, lancet-headed windows. After 1250 the windows became broader, divided up, and ornamented by patterns of tracery, while in the vaults the ribs were multiplied. English Gothic attained its excellence from 1260 to

1290, at which date English sculpture was at its best, and art in painting, coloured glass making, and general craftsmanship was at its zenith.

After 1300 the stone structure began to be overlaid with ornament, the window tracery and vault ribs were of intricate patterns, the pinnacles and spires loaded with crocket and ornament. This later style is known as the "Decorated" and came to an end with the Black Death, which stopped all building for the time. With the changed conditions of life the type of building changed. The "Perpendicular" style—unknown on the Continent—developed with great rapidity and uniformity from 1360 to 1520. As its name implies it is characterised by the perpendicular arrangement of the tracery and panels on walls and windows. It is also distinguished by the flattened arches with square mouldings, by the elaborate vault traceries (especially fan-vaulting), and by the flat roofs and towers without spires.

The medieval styles in England ended with the dissolution of the monasteries (1530–1540), for the Reformation checked the building of churches. Then came the building of the manor house.

The "Tudor" manor was distinguished by flat-headed windows, level ceilings, and panelled rooms. The ornaments of classic style, known as the Jacobean, were introduced under the influences of Renaissance sculpture. At this time too the professional architect arose. Hitherto, building had been entirely in the hands of the builder and the craftsman.

Before the coming of the Normans the study of

building tells us but little of the history of the county. It is true that the monastery at Bangor-is-y-Coed was an important ecclesiastical centre in the fifth century, but nothing remains of this. There were numerous churches scattered over the district and records of these in the eleventh century are found in Domesday Book, but now not a stone is left to guide us as to their construction.

St Asaph Cathedral

Nor is this to be wondered at in a district where war was almost incessantly waged. The buildings were probably of wood and were consequently more liable to fire and destruction. Bede mentions that the church at Lindisfarne was built of sawn timber and wicker, and it is likely that the "Cil" and the "Cor" of the Brython at Cilcain and Bangor and other places were of similar construction.

Flintshire possesses one of the four cathedral churches of Wales. It is situated on the banks of the Elwy at St Asaph, and is supposed to have been founded as early as A.D. 560. The church was burnt down during the wars of Edward I, but it was entirely rebuilt late in the thirteenth century. Owain Glyndwr is accused of having burnt the cathedral to be revenged on Bishop Trevor for publishing the sentence of deposition on Richard II. When Bishop Redman rebuilt it in 1482 the walls only were standing, and possibly the fabric was largely made of wood. In 1638 great trees were used for making a new steeple-loft or belfry, and forty years afterwards the cathedral is recorded to have a "clay flore" and "only a scurvy stone wall in the middle."

St Asaph Cathedral as it stands now is a plain cruciform structure, principally in the Decorated style, with a plain and massive embattled tower rising from the intersection of the nave and transepts. The woodwork of the roof is late Perpendicular. It is the smallest cathedral in England and Wales and probably lost much of its architectural features during the many rebuilding processes it has had to undergo.

There are several interesting old churches in Flintshire. There is a type of parish church peculiar to the county—the "double nave church," two adjacent naves of similar dimensions and structure, so that from the exterior it is impossible to know the position of the chancel. It is sometimes called "the Vale of Clwyd type." The churches at Cilcain, Caerwys, Rhuddlan, Hope, St Asaph, and Llanasa among others are of this design.

Cilcain church has as its most striking feature the beautiful Perpendicular roof of the south aisle, the richest in the diocese. Its western portion is of the "hammer-beam" type, richly ornamented. The eastern portion is covered with horizontal bands of quatrefoil tracery in a circular design. The nave roof was very probably brought from Basingwerk Abbey, after its dissolution.

St Winifred's Chapel, Holywell

Northop church is partly Decorated, partly Perpendicular. Its beautiful tower, 98 feet high, is built in five stages, and forms the most important architectural feature of the church.

Holywell church was re-built in 1769, and copies architecture of various styles. Its tall square tower is early Perpendicular, some of the pillars of the nave are Norman, and the arches under the tower Early

English. St Winifred's Chapel is composed of chancel, nave, and north aisle of three bays. The chancel is of pentagonal shape. The chapel walls are continuous and rest upon the outer walls of the well-chamber, so that the two form structurally one building. The roof of the nave is Perpendicular. It was erected by Margaret,

Mold Parish Church

Countess of Richmond, mother of Henry VII, at the end of the fifteenth century and is a fine example of elaborate Perpendicular work. The celebrated St Winifred's Well is enclosed in a polygonal basin, from the angles of which rise lofty fluted pillars that support the beautiful vaulted roof, the bosses of which are carved with the arms of the Stanley family, Catherine of Aragon, and others.

Mold has the finest church in the diocese. It was built in the reign of Henry VII and consists of a nave with north and south aisles and a western tower, the nave being divided from the aisles by seven arches supported on clustered columns. It is Perpendicular in structure.

Caerwys church is of the "Vale of Clwyd type,"

Caerwys Church

with a massive square tower. Its principal structural features belong to the late Decorated period, but the north aisle shows an interesting example of Early Perpendicular oak roofing, and there are Decorated windows on the south side of the chancel.

Diserth church was largely rebuilt in 1875, but it possesses a beautiful Perpendicular east window of five lights which has the remains of a fine and richly-toned "Tree of Jesse" on glass of the fifteenth century.

Bodelwyddan, known as the "Marble Church," is

Basingwerk Abbey

one of the best examples of modern ecclesiastical architecture in the kingdom. It was built in 1860.

In the year 1132 an abbey was built at Basingwerk, at that time a beautiful part of the coast, and was called "the Monastery of Our Blessed Lady of Basingwerk." It belonged to the Cistercian Order. It is now in ruins, and soon nothing will be left of the once magnificent building. Tradition says that it was founded by Ranulf, Earl of Chester. The church was cruciform, the choir probably being without aisles. The western wall of the refectory has a fine range of Early English arcading. The abbey was built when this period of architecture was at its zenith. At the Dissolution in 1535 its beautifully carved oak and coloured glass were taken to enrich the parish churches in the locality.

At a distance of about half-a-mile from Rhuddlan is Plas Newydd farm house, where once stood a Dominican friary. In the beginning of the eighteenth century its walls stood roofless. But very few remains are now visible and those are mingled with the outer walls of the farm building. On the south side there is a large building 75 feet long, mainly of the late Decorated period, which was probably the hall of the friary. It is used now as a barn. The western wall of the farmyard, nearly 120 feet long, is part of the original building and shows five plain-pointed single-light windows.

Architecture—(*b*) Military.

The earliest form of military architecture in the county is that found in the primitive hill forts, built many centuries ago, probably by the Brythons. They were situated along the south-western boundary on the salient heights of

Sketch Map showing the Chief Castles of Wales and the Border Counties

the Clwydian range, whose escarpments boldly face towards England and form a noble frontier against an invading foe. All are built on the same plan and belong to the late Bronze Age. It was from these that the Brython watched the Roman as he entered his country for the first time. The forts lie in a semi-circular line from Prestatyn to Hope, and among the most important are the following:—

Moel-y-Gaer in the parish of Bodfari. This stands above a pass leading through the Clwydian range to the Vale of Clwyd. There are no defensive works on the eastern side, but remains of an inner and outer bank on the south-east are visible. No Roman remains were discovered in it. It is of oval type and consists of earth and loose shale.

Pen-y-Cloddiau or Bryn-y-Cloddiau ("the Hill of the Enclosures"), a hill about 1400 ft. above Nannerch, shows entrenchments of very considerable importance, forming a large oval-shaped camp, about half-a-mile long, and in parts a quarter of a mile across, surrounded by enormous ditches. This is the largest camp in the county, being quite fifty acres in area. There is another strongly fortified camp of oval shape on Moel Arthur, a circular hill nearly 1500 ft. above sea level lying about a mile to the south-east. It was defended by ditches and dykes of remarkable depth which surround an enclosure of about five acres in area.

Caer Estyn ("the Fortress of Hope") stands on the summit of a pretty hill, Bryn-y-Gaer, overlooking the rocks on which the ruins of Caergwrle Castle stand. It is of considerable size, and, like the other camps, is of an irregular oval shape. It was defended by deep ditches and a well-fortified entrance, and must have been impregnable in prehistoric times. Moel-y-Gaer ("the Hill of the Fortress"), one of the highest points of the Halkyn Mountain, near Rhosesmor, is the site of a large camp. The enclosure is almost circular, with a diameter of 600 ft. Owain Glyndwr and his men occupied it. Bwrdd Rhyfel ("the Plateau of Conflicts") is an old camp in the parish of Ysceifiog, which consists of a simple enclosure of space by a mound and a ditch.

The dykes called by the names of Offa and Wat run across the county from north to south, more or less parallel to each other at a distance varying from 500 yards to three miles. Offa's Dyke begins at the sea near Prestatyn and runs between Llanasa and Gwaenysgor, whence it follows the boundary of Newmarket and Llanasa parishes. It can be traced for a short distance in Whitford parish, and appears for about 100 yards on the north side of Walk Tan-y-Plas, near Caerwys Hall, after which no vestige of it remains till it enters the south-east corner of Tryddyn parish. It then proceeds through Llanfynydd towards Denbighshire, crossing the old Roman villa, Ffrith.

Wat's Dyke begins below Basingwerk Abbey and proceeds through a vast ditch (erroneously called Offa's Dyke) and the Strand Fields, near Holywell, to Cefn-y-Coed along the eastern side of Nant-y-Flint, Coed-y-Llys, Bryn-y-Moel, Mynachlog near Northop, Soughton Park, Llwyn Offa, Bryn-y-Bal, and through Hope parish east of the church, beneath Caer Estyn, ultimately crossing the Alun to Denbighshire.

When the Normans came they built a stronghold at Toot Hill, Rhuddlan, in the form of a moated earthen mound with a wooden stockade. Other castles of the "mount and bailey" type were erected during the Norman period at Trueman's Hill, Hawarden, the Roft on the banks of the Alun in Marford, and Bailey Hill at Mold. Llys Edwin, near Northop, is of later design but constructed on the Norman model. This was the home of Edwin of Tegeingl, from whom many of the Flintshire families claim descent.

The erection of the stone castles at Hawarden and Hope in the middle of the thirteenth century checked the Welsh and prevented them from completely recapturing Englefield. The medieval castle of Caergwrle occupied a commanding position on a hill on the right side of the river Alun, guarding an important passage to Tegeingl. Queen Eleanor, consort of Edward I, stayed here in 1242, when an outbreak of fire partly consumed the interior. It is now a ruin with two towers of Early English design.

Ewloe Castle was built by Llywelyn ap Gruffydd in 1258. It stands on an eminence at the confluence of two streams called Alltami and Northop Brooks in Ewloe woods. It is concealed on three sides and a direct assault of it would have been difficult owing to the depth of its ditches. Its picturesque towers covered with ivy and its stone walls are now in ruins.

There was a mount and bailey castle at Mold, but not a vestige of it remains. It stood on what is now known as Bailey Hill, and covered seven acres of ground. It was built by Robert de Montalt, and though strongly fortified it was taken by Owain Gwynedd in 1147 after a long siege.

Hawarden Castle keep

Castell Ty Maen was situated on a great mound near Gorsedd in the parish of Whitford. Of this traditional residence of Ednowain Bendew, Lord of Tegeingl, the founder of the fifteen tribes of North Wales, no remains are left. A prehistoric stronghold, very probably, occupied the hill on which Hawarden Castle now stands. The Normans took advantage of this admirable site and constructed on it an earth and timber fortress, which was destroyed by the Welsh in 1253. The present castle was built by Edward I, when a round tower about 40 feet high was erected upon a mound, and part of the ground on the south and east side was enclosed within a wall seven feet thick and connected with the tower. Within this enclosure several buildings were constructed, including a chapel, the keep (which still stands in perfect condition), and the hall. One of its fine Early English windows and a portion of its walls still remain. The castle supported the King in the great Civil War and Charles I visited it after the defeat of his force at Rowton Heath.

There was a castle at Basingwerk, but Prince Owain Gwynedd levelled it to the ground in 1165, and of Holywell Castle no traces remain. The same may be said of Prestatyn Castle, which stood on an elevated spot in a meadow not far from Nant Hill. Here, on a low circular mound about sixty feet in diameter, Robert de Banastre built his castle, of which the surrounding ditch is now alone to be seen.

On the summit of a high rock above Diserth stand the remains of Diserth Castle, constructed by Henry III in 1241 and destroyed by Llywelyn in 1263. It now consists merely of shattered fragments of walls and towers ranged round an inner and outer bailey. Unfortunately, even the little that now remains of the walls are soon to be obliterated by blasting operations.

The two most important castles in the county, however, are those of Flint and Rhuddlan.

Flint Castle was built by Edward I in 1277. It was originally of square form, with one of the corners cut off, strengthened by four circular towers. The south-eastern tower stands outside the quadrangle and the sea at high tide surrounded it, but it communicated with the other parts by means of a drawbridge. It is one of the most remarkable towers in Wales, as it consists of two concentric walls each six feet thick, with a space of about 21 feet between them. As it obtains light from the exterior by narrow slits only, it is very dark, and it is difficult to guess what its use was. The inner ward, which measures 160 feet by 145 feet, is now used as a recreation ground, and the Flintshire Territorials have acquired the outer ward as their headquarters.

Rhuddlan Castle was also built in 1277, on a site a furlong northwards of the old castle of Llywelyn ap Seisyllt. Its plan is much like that of Flint Castle, but it is smaller. Edward I and his wife spent many a night in this castle. The ivy-clad walls are about 35 feet high, and had a walk on their summits. The moat surrounded the castle except on the river side, which was defended by a square tower built of red sandstone, possibly all brought from the quarry near Chester. The construction of the castle cost £11,000. This money was raised partly from the revenues of Chester and the Four Cantrevs, partly from three half-yearly payments of Llywelyn's tribute from Anglesey. About £1200 were spent in wages alone when Flint Castle was built. The two castles were probably designed by the same architect.

Architecture—(*c*) Domestic.

Most of the architecture in the county dates after the reign of Queen Elizabeth, but there are at least two pre-Elizabethan buildings, Siamber Wen, now in ruins near Prestatyn, and the Tower near Mold.

Siamber Wen was a stone rectangular mansion of the fourteenth century, built on a T-shaped plan. The middle portion formed a hall which contained a well, the eastern part was the dais for those of higher rank, as in a college hall, and the western part served as a kitchen. There are several slits and windows in the walls for purposes of defence.

The Tower at Mold is a tall, plain, machicolated structure about 27 ft. broad on the south side by 45 ft. on the west. It was built in the fifteenth century, and was inhabited in 1465 by Reinallt ap Gruffydd, one of the six brave captains who defended Harlech Castle in the Wars of the Roses.

The residence of the chief landowner of a district in the Middle Ages, when not a castle, consisted of a hall, usually on the ground floor, and the hall was not only a reception and dining-room, but was also the sleeping place for the greater number of persons living in the house. In many cases there were subsidiary chambers as private apartments for the landowner and his family. As time went on the number of these chambers increased and the importance of the hall diminished, but it impressed itself so firmly on the popular mind that the word still remains in use for the house of the landowner, which is often spoken of as "the Hall."

Of Flintshire mansions Emral Hall in Maelor is cer-

Emral Hall

tainly the most important. The north-west portion of the front was built about the year 1600, and in 1727 extensive additions were made. A moat 168 ft. long and 37 ft. wide extended along the north, west, and south-west sides, and at one time it probably surrounded the house. Several of the rooms have plaster ceilings in elaborate designs of late seventeenth or early eighteenth century date. It was the home of the Pulston family, but Robert Pulston forfeited the estate by joining Owain Glyndwr, his brother-in-law.

There are other fine mansions in Maelor, such as Bettisfield, an old hall of the early seventeenth century, but now converted into a farmhouse. Halghton Hall, a red brick structure of 1662, and Mulsford Hall, built in 1746, with a fine old staircase, a curious inscription, and a

Bryn Iorcyn

moat 168 ft. long on its north-west side. Broughton Hall, a rectangular mansion with two of its sides terminating in four gables, has huge oak beams and finely-carved woodwork and dates from the beginning of the seventeenth century.

The Fferm in the parish of Hope is a quaint Elizabethan stone mansion, the irregularity of its external outline and the style of its chimneys giving a very pleasant effect. Bryn Iorcyn, situated near the summit of Hope Mountain on its east side, was built about 1620. It was once the home of the important Flintshire family of Younge, but the house has been used as a farm for over a century.

Downing Hall in Whitford parish, the home of Pennant, was built in 1627. Its excellent library was sold in 1913, and the curiosities collected by the historian given to various museums.

Mostyn Hall stands in a beautiful and well-wooded park half a mile from the estuary of the Dee. It is built of stone, with mullioned windows and pointed gables, and some of the oldest parts date from the reign of Henry VI. There is a great hall with a dais, and the kitchen is overlooked by a gallery at one end of which is a large room that, tradition says, used to be occupied by Henry Tudor. While the latter was at Mostyn a party attached to Richard III arrived there to seize him, but he escaped through the window, which is to this day called "the King's Window." Mostyn Hall library is the most famous in the Principality, containing many rare editions as well as most valuable Welsh manuscripts. Bychton, in the parish of Whitford, is another manor hall of the same class as Mostyn.

Plas Teg, in Hope parish, is a fine building with square towers of five storeys at each corner of the house. It was erected at the beginning of the seventeenth century, and was once the seat of Sir John Trevor, who was secretary to the Earl of Nottingham, and fought against the Spaniards in the Armada. In the centre is

Downing Hall

a large hall, and above it a dining-room of the same dimensions.

An interesting specimen of old Welsh domestic architecture of the sixteenth century is Pentre Hobyn, in the plain of the Alun between Pont Blyddyn and Mold. It is of yellow sandstone, with pointed gables, mullioned

"Lletty" at Pentre Hobyn

windows, and oak-panelled rooms. Near the house are eight vaulted cells or "lletyau," which were erected as a lodging for travellers after the dissolution of the monasteries, when the monks could no longer entertain strangers.

Golden Grove, near Llanasa, is also Elizabethan, a fine picturesque house built in 1578. Hawarden Castle only dates from the middle of the eighteenth century. There are many other houses of architectural and historical interest in the county, but space will not permit a detailed description of them.

The Flintshire cottages up to the beginning of last century consisted chiefly of two rooms—a kitchen and sleeping apartment, with walls two feet thick made of stones and clay, and a thatched roof. When the burning of lime and quarrying became more common the walls were built with quarried stones and lime and roofed with Llangollen and Carnarvonshire slates. Then brick-making developed into an important industry and two-storey cottages containing four rooms were commonly erected. During the last thirty years there has been a decided improvement in the structure of workmen's dwellings, and artistic and neat cottages are now being built.

19. Communications—Past and Present.

The roads, if there were any in this district before the coming of the Romans, were of the most primitive type. As they were in no way suitable for military purposes, the Romans had to construct others to suit their own requirements. At least one of these crossed our county. This road, starting at Deva (Chester) passed through Hawarden, Northop, and Pentre Halkyn. It then traversed what is now Holywell race-course and passed west of Caerwys towards St Asaph—where slight traces of it remain in an old road at the back of the Bishop's

Palace—crossing the Elwy a little south of the bridge on its way to Segontium. Tradition traces it through Bodfari and Pont Ruffydd, but its true course, as well as the exact situation of the Roman station mentioned by Richard of Cirencester, are still a matter of conjecture. Another Roman road probably connected Chester with the Roman villa at Ffrith in the parish of Llanfynydd.

Bangor-is-y-Coed

The *Via Orientalis*, which connected Uriconium (near Shrewsbury) and Deva, entered Maelor, the detached portion of the county. Faint traces of it are to be found in the parish of Hanmer above Croxdon Pool, and near Willington.

During the four hundred and fifty years from the Norman Conquest to the accession of the Tudors no attempt was made to improve the roads of the county. In 1282 Edward I had to employ great numbers of "hatchet-men" to cut his way through the thick forest that lay between Flint and Rhuddlan. The road was in such a condition in 1685 that it even took the Viceroy—for whose journey the greatest preparations had been

Railway Bridge over the Dee at Shotton

made—five hours to travel the fourteen miles between St Asaph and Conway, and when he arrived at Penmaen Mawr, he had to alight whilst the coach was carried over in portions.

In 1774 Rhuddlan marsh was trackless waste. The only track in the district that could be called a road at this date was that leading from Mold through Holywell,

St Asaph, and Kinmel Park over Penmaenbach towards Bangor. People travelled on horseback, and cartage was almost impossible. Traffic had to be done by pack-horses, mules, and oxen. A horse's load was usually three hundredweight, or four bushels of grain. Two-and-a-half miles an hour was considered a fair rate of progress. When the farmers began to use waggons they had to send in advance a boy blowing a large horn, to give notice to any other vehicle approaching to halt at a place where the two could pass each other.

In 1769 the Highway Act was passed, and a four-in-hand coach began to run, carrying mails and passengers. The stage coach highway in this county ran through Mold, Pentre Halkyn, Holywell (a halting station), Gorsedd, *via* Travellers' Inn, to St Asaph and Kinmel.

In 1836 the Government determined to construct a road across Rhuddlan marsh in a straight line from Rhuddlan to Abergele. To defray part of the cost of this road a number of toll-gates were placed at intervals along it. Toll-gates were first constructed about the middle of the eighteenth century. The "toll" has now been abolished, and the cost of repair is borne by the County Council.

To-day three principal roads cross the county lengthwise, inclining in a north-westerly direction—one runs through Caergwrle to Mold and Nannerch in the Alun Valley, thence to Bodfari along the beautiful Vale of the Wheeler; another passes through Hawarden, Northop, and Holywell over the Clwydian range on its way to St Asaph and Bodelwyddan; and the third is the flat coast road from Saltney through Flint, Mostyn, and Prestatyn. For about two miles from Saltney it is part of the "old road" and its meandering course followed the coast line before the "cut" was made for the Dee. These three, as the map shows, are intersected by a network of roads in every direction from one end of the county to the other—one interesting branch being that which turns north-west at Lloc to Rhuddlan *via* Hiraddug Pass, where it takes the name of "Ffordd Ffrainc"—the Frenchman's (i.e. the Norman's) Road.

The first meeting relative to the construction of the London and North Western Railway from Chester to Holyhead was convened in London May 1st, 1839. Many and formidable objections were raised against the railway scheme, especially by the farmers, who contended that their stock would be unduly alarmed, their property destroyed, their labour disorganised, and their stockyards inevitably set ablaze by sparks emanating from the locomotives. However, in 1845, the first sod of the line from Chester to Conway was cut. This section cost on an average £22,000 a mile. According to the first time-table, eight trains a day—four each way—were run. In 1854 the Vale of Clwyd section of the railway from Rhyl to Denbigh was opened. In about twenty years this railway was leased to and worked by the London and North Western Railway Company. Another line was opened along the Vale of the Wheeler from Mold to Denbigh which was also taken over by the same company, and later extended to Chester.

The Coed Talon and Brymbo Railway, under the

control of the same Company runs through a district famous for its romantic scenery from Mold, passing through Llanfynydd and Ffrith to Brymbo.

In 1912 the picturesque railway from Greenfield to Holywell was opened for passengers, and a scheme is now on foot for making one from Holywell to Halkyn, as this rich mining district is seriously handicapped without a railway.

A branch railway, familiar to tourists, runs along the foot of the mountains from Prestatyn to Diserth, and will be shortly extended to Newmarket. The Great Central Railway runs across the eastern portion of the county through Shotton, Hawarden, and Caergwrle, to Wrexham, and thence proceeds to Bangor-is-y-Coed and Overton. The Cambrian Railway also serves the Maelor district, passing through Fenn's Bank and Bettisfield, and the Great Western Railway route to London enters the detached portion Marford and Hoseley.

A mineral railway runs from Connah's Quay to Buckley catering for the numerous coal mines and brick works in this populous district, under the working control of the Great Central Railway.

The Shropshire Union Canal serves the southern end of Maelor, and runs parallel with the south-eastern boundary on the Shropshire side.

20. Divisions, Ancient and Modern. Administration: Population.

Wales was divided in the tenth century into three main divisions—Gwynedd, Powys, and Dinefawr. Modern Flintshire forms part of Gwynedd and part of Powys.

Gwynedd was again divided into four parts. Of these Y Berfeddwlad (called thus, "the Middle Country," because it was situated between Powys and Gwynedd), was one.

The Berfeddwlad contained five cantrevs and 13 commotes. The cantrev was a division which probably represented one hundred free families, or one which could produce a hundred warriors to defend its land. One of these cantrevs was Tegeingl, which is to-day included in the county of Flint, and its three commotes are called Rhuddlan, Prestatyn, and Cynsyllt. It is also probable that a portion of Llannerch, a commote in Dyffryn Clwyd, penetrated Flintshire near Bodfari.

The south-east of Flintshire and Maelor Saesneg beyond the Dee, as well as Bromfield on the Alun, Caergwrle, Estyn (Hope) and the Mold district, were in Powys Fadog.

The commote was the unit of the government and administration. It was governed by a lord, a man of noble blood, and his chief officers were the "maer" and "canghellor." It had its own court of justice, and closely resembled the English manor.

The parish is an ancient division which was originally

a "township" or a cluster of houses to which a priest, to whom its tithes were paid, ministered. In the reign of Queen Elizabeth the parishes were made areas for taxation, partly to provide funds for the relief of the poor. In modern times, the object being better care of the poor, the parishes have been grouped together in the Poor Law Unions, of which there are four in Flintshire, each provided with a workhouse, which was meant originally to be a place where the able-bodied poor might find employment. Parish affairs since 1894 have been managed by Parish Councils, which are composed of representatives elected by the parishioners.

The District Council has charge of the wider areas, and is responsible for the sanitation and the water supply of the district. The County Council is, however, the main governing body of the county, having power to levy rates and borrow money for public work. It manages the lunatic asylums and reformatories, keeps roads and bridges in repair, controls the police, appoints coroners and officers of health, and sees that the Acts relating to local government are carried out. It has the management of the education of the county, which it relegates to the Education Committee. It resembles the ancient Shire Moot or county assembly, and has its headquarters at the County Offices at Mold.

The Flintshire County Council has 56 members, of whom 42 are elected every three years to represent the various districts, while 14 are aldermen co-opted by the Council, seven for the term of three years, seven for the term of six years.

The chief officials of the county are the Lord-Lieutenant and the High Sheriff. The Lord-Lieutenant was the Custos Rotulorum—the keeper of the rolls—i.e. the records of the County Sessions; but these are now kept by the Clerk of the Peace. The office of the Lord-Lieutenant dates from the fourteenth century. He was then the chief military officer of the Crown in the

Town Hall, Mold

county. He selects the county magistrates, who are appointed by the Lord Chancellor. He is appointed for life.

The Sheriff was originally elected by the people of the county, but from the time of Edward II he has been appointed by the Crown "on the morrow of the St Martin's, November 12th." As representative of the

King he collects the King's debts, and is the military as well as the judicial executive head of the county. He is the first man in the shire, taking precedence over all peers and the Lord-Lieutenant. He presides over the election of members of Parliament.

The court-leet, which meets in several Flintshire manors, is a survival of the old court of the manor, where the lord met his tenants and arranged local matters, such as the right to the common land and the holding of fairs and markets.

Flintshire is now divided, like other counties, into the Ancient or Geographical County, the Administrative County, and the Registration County. The Ancient or Geographical County, as formed by Henry VIII, contains an area of 164,744 statute acres. This again is divided into hundreds, parishes, townships, Poor Law unions, and sanitary districts. The Administrative County, that is, the area over which the county authorities exercise control, contains 163,025 statute acres, and includes Sessions, County Court Districts, Councils for administration of Justice, Borough and County Constituencies for the appointment of Parliamentary representatives, County Council, and Urban, District, and Parish Councils. The Registration County, which represents the area as divided for Parliamentary purposes, contains an area of 74,023 statute acres.

The Administrative County has a population of 92,705 in 1911, as against 81,485 in 1901, an increase of 13·6 per cent. The Registration County in 1901 contained a population of 42,261. The disparity between the population of the Registration County and that of the Administrative County is due to inclusion of several parishes in the latter which for registration purposes were included in the other neighbouring counties; the popula-

Town Hall and Carnegie Library, Rhyl

tion thus included in Cheshire was 18,275, in Denbighshire 15,892, and in Shropshire 5057.

In 1801 the population of the Geographical County was 39,469; in 1901 it was 81,485. Flint has, therefore, rather more than doubled in population during the century. The Ancient County is not divided for Parliamentary purposes. It contains the Flint district of boroughs, comprising the contributory boroughs of Flint, Rhuddlan, Caerwys, Caergwrle and Overton, together with those of Holywell, St Asaph, and Mold, which were added by the Reform Act of 1832; the only town exercising all the powers of a borough in the county is Flint. These send a representative to Parliament, and a member is also elected for the county. The population of these contributory boroughs in 1901 was as follows: Caergwrle, 1345; Caerwys, 526; Flint, 4661; Holywell, 6873; Mold, 4263; Overton, 1111; Rhuddlan, 997; and St Asaph, 1765.

There are 57 ecclesiastical parishes or districts situated wholly or partly within the Ancient County. Of these parishes or parts of parishes, fifty-three, with an aggregate population of 79,304, are within the diocese of St Asaph; three, with a population of 2077, are within the diocese of Chester; and one, with a population of 319, in that of Lichfield.

The Administrative County has one court of Quarter Sessions, and is divided into nine Petty Sessional divisions—Caerwys, Hawarden, Holywell, Hope, Mold, Northop, Overton, Prestatyn, and Rhuddlan, each having magistrates or Justices of the Peace, who attend to try cases and punish petty offences against the law. The municipal borough of Flint has a separate commission of the peace, but no separate court of Quarter Sessions. The business of Quarter Sessions since the establishment of the County Council in 1888 has been considerably reduced and is now confined to criminal business, and the licensing of the public houses.

The total population in 1911 of the Urban Districts in the Administrative County (comprising the municipal borough of Flint, and six other Urban Districts) is 34,864, an increase of 4414 or 11·2 per cent. since 1901. The county contains four rural districts, the aggregate population of which is 57,841, an increase of 6446 or 12·6 per cent. since 1901. The rural districts are those of Hawarden, Holywell, Overton, and St Asaph.

Flintshire is divided into three Poor Law Unions: Holywell, Hawarden, and St Asaph, and in addition a portion of Whitchurch and Ellsemere Unions.

21. Roll of Honour.

Flintshire, though of small area, has a long roll of distinguished men. In border counties we may often remark the existence of great families which seem to have come into prominence through the very stress of the constant turmoil and fighting, and to hold their position over a long period of time. Such is the case in Flintshire. It was the home of the Stanleys, who played such an important part in the Wars of the Roses, though not always with conspicuous fidelity, for while during the early part of the war they sided with Richard III they turned over to the Lancastrian party at the critical moment at Bosworth Field. Lord Stanley and his brother had brought 3000 men to the battlefield, but as Richard's army melted away they joined the standard of Henry Tudor.

The Mostyn family, who have figured prominently in the history of the county up to the present day, had also much to do with the issue of events at Bosworth Field, for one of them, Hugh Conway of Bodrhyddan, went to Brittany to acquaint Henry of the intended rising in his favour, and Richard ap Howell of Mostyn led 1600 men to the fight on his side and received from Henry the belt and sword he was wearing in reward for his aid.

The Glynnes and Trevors also have long been well known in connection with the county. Sir John Glynne espoused the Parliamentarian cause against Charles I, and purchased the manor and the rectory of Hawarden from the agents of sequestration, though from the time of Henry VI they had been in the possession of the Stanley family. Sir John Trevor of Trevalun was secretary to the Earl of Nottingham, Comptroller to the Navy in the time of Queen Elizabeth and James I, and was head of a powerful family holding several seats in Flintshire and representing the influential clan of Tudor Trevawr of Hereford.

Our county has been the birth-place of a number of historians. Ellis Griffith, who described himself as "A Soldier of Calais," was a native of Gronant Uchaf, who left Wales about 1528, joined the army, and was stationed at Calais for some years. He wrote the history of England and Wales from the time of William the Conqueror to Edward VI, a work of considerable importance which is one of the most valuable manuscripts in the

Mostyn Library. John Jones, of Gelli Lyfdy, Ysceifiog, who lived in the middle of the seventeenth century, made Welsh history his special study, collecting a valuable

Thomas Pennant

library of books on the subject and transcribing old Welsh manuscripts. A large number of his books are in the Mostyn Library. To Thomas Jones of Denbigh (1756–1820), who was born at Caerwys, Wales is much indebted for his services to the literature of the Principality. He published numerous books, and wrote many hymns, but his best-known work was his History of the Reformers.

All these names, however, must yield to that of Thomas Pennant (1726–1798) of Downing, near Holywell, who as historian, zoologist, traveller, and antiquary left behind a vast mass of information and a European reputation as a naturalist. He wrote many books, was a Fellow of the Royal Society and the friend of Linnaeus, and was a man of unusually fine character. His *Tours in Wales* and the *History of Holywell and Whitford* are the works which are of most interest to Welshmen. He was buried at Whitford.

Among writers who have more particularly distinguished themselves are Thomas Edwards, "Caerfallwch," born near Mold in 1779, the compiler of a Welsh and English dictionary; the Rev. John Williams (1809–1862), founder, with Longueville Jones, of the *Archaeologia Cambrensis*, and editor of *Annales Cambriae* and the *Brut y Tywysogion* or the Chronicle of the Welsh Princes; Roger Edwards (1811–1886) who spent 40 years at Mold, starting the leading review of Wales, *Y Traethodydd*, and editing *Y Drysorfa*; and Daniel Owen, the Dickens of Wales, whose novels are known to most Welsh readers. His works are remarkable for their pathos and humour and their popularity is evinced by the bronze statue erected to his memory at Mold by public subscription.

Matthew Henry (1662–1714) was born at Broad Oak in Iscoed Maelor, the son of one of the 2000 clergy who left the Church of England in consequence of Charles II's

Act of Uniformity. He became one of the greatest Biblical scholars of his day and published his famous *Exposition of the Old and New Testament* in 1710, though complete only to the end of Acts. Another well-known Biblical scholar was Dr Parry (1560–1623), Bishop of St Asaph. Dr Morgan had considerably improved upon Salisbury's translation of the New Testament, but Parry improved on both, and thus rendered lasting service to his fellow-countrymen.

The county is not without poets. Dafydd ap Edmund was a native of Hanmer in Maelor, and lived on his estate at Pwll Gwepre. At the famous Eisteddfod held at Carmarthen in 1451 under Gruffydd ap Nicholas he won the chair, and the sanction of the Eisteddfod was given to the twenty-four new canons of poetry which he, with the assistance of the other bards of North Wales, had compiled. John Blackwell, "Alun," who was born the son of a collier at Pont Erwyl, Mold, in 1797, and died in 1840, has been described as "the sweetest and most finished of Welsh poets." He was made curate of Holywell in 1827. It was in this town, too, that Thomas Lloyd Jones, "Gwenffrwd," was born, whose untimely death from yellow fever at the age of 24 removed one of the most promising poets of the early part of the nineteenth century. He died in America in 1834.

Flintshire apparently can claim no artist of special distinction, but in the realm of music must be recorded the name of Edith Wynne (1840–1885) the ballad singer, who was born at Holywell and trained in Florence. She went on an extended and most successful tour in the

United States with Madame Patti in 1871, and was a very popular singer in England about this period. John Ambrose Lloyd, of Mold (1815–1874), a composer of

William Ewart Gladstone

merit who wrote anthems and other church music, also deserves mention.

Perhaps the most distinguished person associated with our county was William Ewart Gladstone, who, though a Liverpool man, was both by marriage and life-long residence connected with Hawarden. Gladstone's career being essentially political needs no mention here, but

Henry Morton Stanley

he was no less famous as a brilliant scholar, and probably one of the greatest English orators of modern times. Born in 1809, he entered Parliament for Newark in 1832, and was four times Prime Minister. He died May 19, 1898, and was buried in Westminster Abbey.

Henry Morton Stanley (otherwise John Rowlands), the traveller, was born at Denbigh, but educated at St Asaph workhouse, where his autograph is yet to be seen. From here he went to relatives at Brynford where he acted as pupil teacher. When about fifteen he set out for Liverpool, and sailed as cabin boy to New Orleans, where he was employed by a kindly master whose name of Stanley he assumed, but on his death he had to face the world penniless. His adventures during the American War brought him before the public as the principal journalist of the United States. In 1868, during the Abyssinian War, he acted as correspondent of the *New York Herald*. At the close of that war he visited Persia and India. In 1871 he went in search of Livingstone, from whom the outer world had received no tidings for a long period, and setting out from Zanzibar made his memorable march through Africa which resulted in his meeting with Livingstone. After various other work in Africa, including the exploration of the river Congo, he was despatched in March, 1887, with a large expedition, in search of Emin Pasha, and, after long delay and hardship, eventually reached the East Coast of Africa in December, 1889. He died in 1904.

22. THE CHIEF TOWNS AND VILLAGES OF FLINTSHIRE

(The figures in brackets after each name give the population in District or Parish in 1911, and those at the end of each section are references to pages in the text.)

St Asaph (1833), or Llanelwy, the smallest city in Britain, is beautifully situated on an elevation between the confluence of the Clwyd and the Elwy, almost in the middle of the Vale of Clwyd, and is one of the most charming places in North Wales. In the sixteenth century it was famous as a place of pilgrimage to the image of Derfel Gadarn, the offerings consisting of horses, cattle, and silver. The city is the site of one of the four Welsh cathedrals, the smallest of all our cathedrals. In front of the cathedral stands a handsome monument, a tercentenary memorial of the translation of the Bible into Welsh by Bishop Morgan which was erected by public subscription.

In the burial ground of the Welsh church, near the wall, is the tomb of Dick Aberdaron, the eccentric Welsh linguist, who is said to have taught himself 30 languages. Traces of the old Roman road are found in the city. Pont Dafydd, the stone bridge of one arch that crosses the Clwyd, was built in 1630.

The cathedral was first erected to St Kentigern in the fifth century. Asaph was Kentigern's successor. Among its bishops was William Morgan, the eminent linguist and translator of the Bible into Welsh. (pp. 10, 14, 21, 51, 58, 64, 98, 105, 106, 113, 118, 119, 137, 139, 140, 148, 149, 153, 156.)

Bagillt is a large village on the estuary of the Dee between Flint and Holywell. It was an important harbour before the silting up of the river. Its chief industries are the lead-works, where the lead ore from Halkyn mountains is treated, and its extensive coal mine, Bettisfield Colliery, which is one of the largest in the county. To the west stand the ruins of Basingwerk Abbey, to the south the site of the battle of Coleshill. (pp. 41, 63, 65, 81.)

Bangor on the Dee (574) or **Bangor-is-y-Coed**, "the High Choir in the Wood," in the detached portion of the county, five miles from Wrexham and 13 miles from Chester, is pleasantly situated on the river Dee, which is here crossed by a handsome bridge of five arches. Its former magnificence has, however, totally disappeared. The place is described by William of Malmesbury, shortly after the Norman Conquest, as consisting of numerous demolished churches and mutilated ruins. Leland, in 1550, says that the site of the old monastery lay in a fertile valley on the south side of the Dee, but that the river had changed its course and at that date ran through the middle of the ground on which it stood. Such place-names as Street Lydan (Broadway), Porth Wgan (High Gate) indicate the former size of the place. The ancient monastery contained 2400 members, who were employed alternately in study, prayer, and labour; 100 being constantly employed in divine worship, one band relieving the other every twenty-four hours. Bede tells us that over 1000 Bangor monks, who prayed for the success of the Welsh at the battle of Chester, were killed by Ethelfrith. (pp. 83, 113, 115, 118, 142.)

Bettisfield (356), a township in the parish of Hanmer, stands on the borders of Shropshire. Bettisfield Park was the birthplace of Sir Thomas Hanmer, Speaker of the House of Commons in the reign of Queen Anne. (pp. 133, 142.)

Bodfari (322), four miles N.E. of Denbigh, an ancient village on the river Wheeler on the east side of the Vale of Clwyd, is supposed, though doubtfully, to have been the Roman station "Varis." About three hundred yards from the church is Deifer's Well, with which many curious customs are associated. (pp. 19, 28, 65, 113, 115, 125, 137, 140, 143.)

Buckley (6333), a very populous village partly in the parish of Hawarden and partly in that of Mold. It is the "Biscopestreu" of Domesday Book. It has important potteries for the manufacture of coarse earthenware, fire-bricks, drain-pipes, and tiles. There are also large collieries in the district, some of which have been worked for over six centuries. (pp. 19, 40, 41, 54, 64, 65, 105, 142.)

Caergwrle (1345), five miles N.W. of Wrexham, is a quaint little town on the banks of the Alun and one of the Flint District Boroughs. Ruins of a picturesque thirteenth century castle still exist. It is an inland watering-place and has sulphur and saline wells of some repute. There are traces of Roman roads here. (pp. 11, 15, 16, 24, 112, 113, 115, 125, 127, 140, 142, 143, 148.)

Caerwys (834), 7½ miles E. of St Asaph, is a contributory borough. It was formerly of considerable importance, as the assizes were held here until the middle of the seventeenth century, when Flint became the place of judicature for the county. The old gaol and the market hall have been converted into dwelling houses. In the wood below Maes Mynan once stood the seat of Llywelyn, the last Welsh prince, his traditional resort in time of peace. Caerwys was renowned for its Eisteddfodau. Gruffydd ap Cynan held an Eisteddfod here in 1100, and another was held in 1568, under a commission from Queen Elizabeth. The parish contains numerous wells, tumuli, and prehistoric remains. Offa's Dyke appears near Caerwys Hall. (pp. 11, 19, 28, 35, 40, 80, 90, 110, 111, 113, 115, 119, 122, 126, 127, 137, 148, 151.)

Cilcain (787), a village on the side of Moel Fammau, six miles S.W. of Flint. The church has a finely carved roof. There are tumuli in Coed Du and an interesting stone circle in Pen Bedw Park. (pp. 23, 27, 28, 35, 95, 115, 118, 119, 120.)

Connah's Quay (4596), a town of modern growth, four

miles S.E. of Flint, in a very populous district. It is one of the most important ports on the estuary of the Dee and has a considerable shipping trade. (pp. 54, 64, 88, 89, 142.)

Diserth—"the deserted place"—(902) is a beautiful village nestling at the foot of Ochr-y-Foel, with ruins of a fine castle. There are valuable quarries of limestone in the vicinity as well as the Talargoch lead mines of Roman fame. The church has a historic cross, "Croes Einion." (pp. 14, 27, 36, 43, 51, 65, 80, 95, 96, 115, 122, 129, 133, 142.)

Diserth Village

Ffrith, a small village near Wrexham in the parish ot Llantynydd. Recent discoveries prove that it was once a Roman villa. Offa's Dyke passes through the village. (pp. 25, 90, 113, 126, 138, 142.)

Flint (5472), one of the eight boroughs of the county, was at one time the county town. It is on the estuary of the Dee and was once an important harbour, but the tide no longer washes the castle walls. The County Hall and gaol have been transformed into houses. Edward I made a grant of a market and fair to the town, which owes its origin to its castle, built in 1277. It was laid out as a regular parallelogram, and surrounded by a ditch, which can still be traced at intervals—in Earl Street, Evan Street, Coleshill Street, and Chapel Street. It was here that David, Llywelyn's brother, made a treaty with Edward I to

Gwaenysgor Font

fight against his own people, and in its castle Richard II met Henry Bolingbroke, and left for London, only to be deposed and to meet with an untimely death. The removal of the assizes to Mold does not deprive Flint of the title—technically speaking—of County Town. There are numerous thriving industries in the town and district. (pp. 2, 6, 8, 12, 40, 54, 64, 65, 72, 87, 89, 90, 97, 101, 102, 103, 105, 106, 107, 108, 115, 130, 131, 139, 141, 148.)

Greenfield (Maes Glas), a populous village on the estuary of the Dee, $1\frac{1}{2}$ miles from Holywell. It is famous in history as the site of "Dinas Basing" and Basingwerk Abbey, and as the termination of Wat's Dyke. Before the construction of the L. and N.W. Railway it was an important harbour. Paper-works and collieries are its chief industries. (pp. 71, 89, 142.)

Gronant and **Gwespyr** are villages situated about two miles east of Prestatyn, famous for their sandstone and whetstone quarries. (pp. 37, 40, 49, 65, 80, 96, 150.)

Hawarden Castle

Gwaenysgor (211) is a hamlet near Newmarket, set on the hills above Meliden. There are several tumuli in the neighbourhood, and it is thought to have been at one time a settlement of the Vikings. The termination *-ysgor* may be Danish. The church has a late Norman font. Ffynnon Wen—the Blessed Well—is in the parish. (pp. 64, 65, 96, 126.)

Halkyn (1339) (Helygain = willow tree) is a village on the east side of Halkyn mountain, three miles S.W. of Flint. Its extensive lead mines, which have given employment to many people for generations, are in a flourishing condition, and there are marble and chert quarries near the village. (pp. 6, 17, 19, 27, 35, 38, 44, 47, 54, 59, 64, 78, 79, 81, 82, 126, 137, 140, 142.)

Hanmer (408), is a village in the detached portion of the county. It takes its name from a spacious mere or lake, on the road between Overton and Whitchurch. The situation of the village is strikingly beautiful. The church is a handsome Early English building, with a lofty square tower. Owain Glyndwr was probablv married in it. (pp. 27, 28, 138, 153.)

Hawarden (6490), a small town lying five miles due S. of Chester, consists of a single street more than half a mile long. The Welsh name is Pen-ar-lag ("Headland of the Lake"), the surrounding district having been once covered by the sea. Large collieries and brick and tile works are in the neighbourhood. Its castle, now in ruins, played an important part in the history of the county. The town is famous as the home of William Ewart Gladstone, four times Premier of England. St Deiniol's Library was erected from the proceeds of the National Memorial to Mr Gladstone. (pp. 14, 18, 40, 41, 58, 64, 65, 71, 94, 96, 98, 100, 101, 103, 105, 107, 108, 127, 128, 129, 136, 137, 140, 142, 148, 149, 150, 155.)

Holywell or Treffynnon (2549) takes its name from its famous well, the largest spring in Great Britain. It is pleasantly situated on a hill overlooking the estuary of the Dee. When the lead mines were at their best and the numerous industries along St Winifred's stream were flourishing, Holywell was one of the busiest towns in North Wales. The church is a fine specimen of Norman and Early English architecture, and St Winifred's Chapel, built by Margaret, Countess of Derby, the mother of Henry VII, is a beautiful example of the Perpendicular period.

Owing to its dry invigorating air the town is a growing health resort. The chief places of interest are the Well, Basingwerk Abbey, Wat's Dyke and Pantasaph Monastery. Holywell is the birth-place of the singer Edith Wynne and the poet Gwenffrwd. (pp. 14, 15, 17, 25, 28, 35, 36, 38, 53, 54, 64, 66, 70, 71, 115, 120, 126, 129, 137, 139, 140, 142, 148, 149, 152, 153.)

Hope (4806), a small village four miles S.E. of Mold stands on a gentle rise on the northern side of the Alun. It is also called Estyn and Yr Hob, and it is mentioned in the Domesday Book. It formed part of Denbighshire in the reign of Henry VIII. The village overlooks Caergwrle and its castle. (pp. 11, 14, 18, 19, 55, 65, 101, 103, 104, 115, 119, 125, 126, 127, 133, 134, 143, 148.)

Leeswood (Coed Llai), a village near Mold, is famous for its collieries. Cannel coal was once found in large quantities, but the seam is now exhausted. (pp. 35, 40, 41.)

Llanasa (3026) is a village in a pleasant valley at the northern extremity of the county. In the olden times it formed part of the estates of the Countess of Richmond. The church is dedicated to the patron saint and founder of St Asaph, A.D. 513. Near it are many other places associated with this saint's name, such as Pant Asa (Asa's Hollow), Ffynnon Asa (Asa's Well), Onnen Asa (Asa's Ash). (pp. 37, 55, 119, 126, 136.)

A menhir (*maen hir*) lies in the yard of the Red Lion inn.

Meliden (560) (Gallt Melyd), a village between Diserth and Prestatyn, mentioned in Domesday as the "Villa of Melyd," was a thriving place when the Talargoch lead mines were being extensively worked. (pp. 27, 51, 65.)

Mold (4873) (Yr Wyddgrug) is built on a gentle acclivity in a fertile plain watered by the river Alun. It is practically the county town, for the assizes and quarter sessions and County Council meetings are held here. Mold district was added to this county in the reign of Henry VIII. Tradition says that at Maes-y-Garmon, near the town, the Bishops Germanus and Lupus defeated the Picts and Saxons in the so-called "Alleluia Victory." Not a stone of the castle remains. The Tower is one of the few pre-Elizabethan houses in the county, and the church, which is mainly of the Perpendicular period, is the finest in the Diocese. The grave of Richard Wilson, the landscape painter, is in the churchyard. Daniel Owen, the Welsh novelist, was a native of Mold; his statue stands in front of the county buildings. The district is particularly rich in minerals and coal. (pp. 11, 14, 18, 19, 24, 28, 40, 41, 55, 59, 63, 65, 97, 98, 99, 100, 101, 105, 108, 112, 121, 122, 127, 131, 136, 139, 140, 141, 142, 143, 144, 148, 152, 155.)

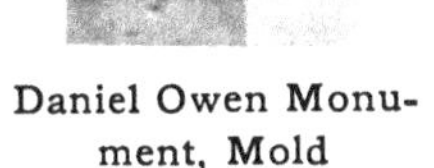

Daniel Owen Monument, Mold

Mostyn and **Ffynnon Groew** are two villages on the estuary of the Dee, about five miles N.W. from Holywell. The inhabitants are mostly employed in the ironworks and the coal mines. The coal mines in this locality are worked for nearly a mile under the sea. (pp. 40, 41, 46, 49, 53, 54, 64, 65, 69, 73, 83, 88, 89, 107, 134, 141, 150.)

Nannerch (300), a picturesque little village on an elevated portion in the beautiful Mold and Caerwys valley, six miles N.W. of Mold. There are two prehistoric camps in the parish, Moel

Arthur and Pen-y-Cloddiau. (pp. 15, 19, 23, 28, 55, 112, 125, 140.)

Newmarket (438), is a village standing on the high table-land above Diserth on the road from Chester to Rhuddlan. Near the village is a hill called Gop-y-Goleuni, on the summit of which is an enormous cairn formed of limestone—the largest in all Wales—the traditional grave of Queen Boadicea (Buddug). It covers nearly an acre of ground and its height is about 80 feet. Below the cairn is a cave which contained numerous prehistoric remains. (pp. 14, 25, 28, 45, 65, 110, 115, 126, 142.)

Northop (2809) (North Hope) is a pretty village lying in a fertile part of the county, 3½ miles S. of Flint. Moel Gaer, which was once occupied by Owain Glyndwr, and Llys Edwin, the home of the Welsh chieftain, Edwin of Tegeingl, are situated in this parish. Its Welsh name is Llanewrgain. (pp. 40, 59, 120, 126, 127, 137, 140, 148.)

Overton (1196), a neat little village on the banks of the Dee on the road from Chester to Shrewsbury. Its handsome stone bridge of two lofty arches connects the counties of Denbigh and Flint. It is one of the Flint boroughs and was a market town. It has the remains of a castle and an Early English church. The numerous yew trees in the churchyard are remarkably beautiful. (pp. 115, 142, 148.)

Prestatyn (2036), a rapidly growing watering-place on the Chester and Holyhead Railway, is situated four miles E. of Rhyl. The name was given to the district by the Mercians about 800. The old castle and the ruins of the pre-Elizabethan mansion, Siamber Wen, are of historical interest. Owing to its bracing and dry atmosphere, its even temperature and low rainfall, it is famous as a health resort. (pp. 2, 7, 14, 18, 35, 49, 50, 51, 64, 87, 92, 96, 98, 126, 129, 131, 141, 142, 143, 148.)

Queensferry is a growing village on the Dee estuary. Its old name was King's Ferry, but out of respect to Queen Victoria, who was ferried across here in early days, and by special request, the name was changed to Queen's Ferry. (pp. 40, 72.)

Rhuddlan (1607), a town of great antiquity, stands on an eminence on the eastern side of the Clwyd, about three miles from its mouth. It was formerly a place of considerable magnitude and importance but few traces of its former grandeur remain, save the ruins of its castle. Edward I gave it the privileges of a free borough. It is famous in history on account of the great battle fought on Rhuddlan marsh, where Offa and his Saxons defeated the Welsh under Caradoc in A.D. 795. The "Statute of Wales" was enacted here by Edward I in 1284. (pp. 1, 2, 10, 21, 51, 52, 55, 64, 65, 87, 89, 91, 92, 93, 94, 95, 96, 97, 98, 99, 100, 101, 102, 103, 104, 105, 108, 115, 119, 124, 127, 130, 139, 140, 141, 143, 148.)

Rhyl (9005), a very favourite modern health resort, stands at the mouth of the Clwyd at the end of the Vale of Clwyd, the Paradise of Wales. It is the largest town in the county and is increasing rapidly. In 1833 it had but two fishermen's cottages, but at the last census its population was 9005. (pp. 4, 9, 10, 19, 46, 47, 51, 52, 59, 60, 64, 65, 87, 88, 141.)

Saltney and **Sandycroft** on the Dee estuary, are extensive villages lately developed on account of their contiguity to Chester and the numerous industries in the locality. The population in 1901 was 3727, in 1911, 6594—an increase of 2867. Before the "cut" was made for the Dee the land was a marshy waste and the fish and rushes belonged to the monks of Basingwerk Abbey. Here Henry II encamped before the defeat of his army in Ewloe Woods. (pp. 4, 53, 54, 65, 72, 97, 141.)

Sealand (755) is that part of Flintshire which is on the Cheshire side of the river Dee, and its rich alluvial soil is famous for its fertility. It is approached by a bridge erected in 1895 at the cost of £23,000. On a charming spot once covered by the sea a "garden city" is now being established. (pp. 20, 53.)

Shotton is a large modern village between Connah's Quay and Queensferry on the London and North Western and Great Central Railways. Hawarden Bridge Iron Works, in the vicinity, employ over 5000 hands. In time gone by towns were built near a castle for protection, e.g. Flint, now it is the industrial centres that account for many a village and town, like Shotton. (pp. 105, 139, 142.)

Tremeirchion or **Dymeirchion** (604) is a hamlet three miles S.E. of St Asaph, situated on the hills in a beautiful country on the eastern side of the Vale of Clwyd. It is a place of great antiquity, famous for its caves and wells. St Beuno's College stands near the village. (pp. 14, 28, 46, 60, 62, 111, 115.)

Whitford (3219) (Rhydwen), mentioned in Domesday Book as Huitford, is one of the prettiest villages in the county, three miles N.W. of Holywell. There are several manor houses in the parish, Mostyn, Downing (the home of the historian Pennant), Bychton, Mertyn, and Y Gelli, as well as many other objects of interest to the antiquarian—Maen Chwyfan, the Tower on Garreg (a sixteenth century beacon), Y Gorseddau, and an inscribed stone at Downing. (pp. 28, 55, 78, 79, 92, 96, 113, 126, 128, 134.)

Willington (308), a village in the detached portion of the county, is celebrated on account of its ancient cross. (p. 138.)

Worthenbury (431), an ancient village, six miles S.E. of Wrexham, mentioned in Domesday Book, stands close to the river Dee. There are very old mansions in the parish, Emral Hall, Mulsford Hall, and Broughton Hall.

Ysceifiog (1121), a village four miles S.W. of Holywell, mentioned in Domesday Book, is situated on an elevated table-land above the Mold and Caerwys road. The church is partly Norman, with a lofty tower. William Edwards ("Will Ysceifiog"), the poet, is buried in the churchyard. (pp. 6, 7, 23, 27, 35, 55, 80, 85, 86, 95, 105, 115, 126, 151.)

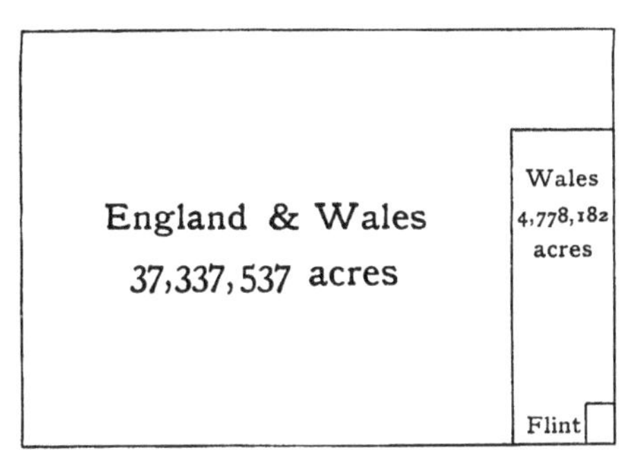

Fig. 1. Area of Flintshire (163,025 acres) compared with that of England and Wales

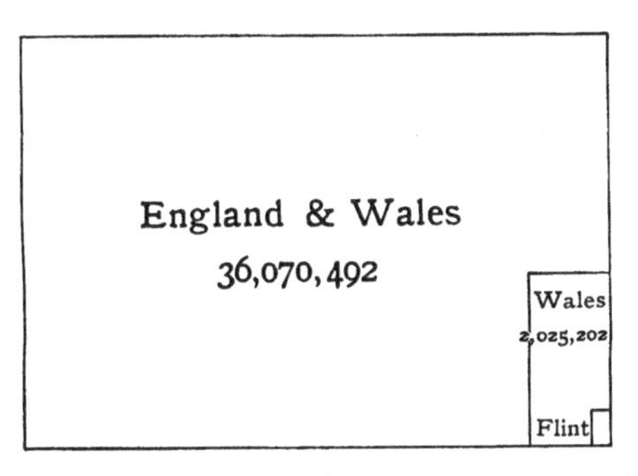

Fig. 2. Population of Flintshire (92,705) compared with that of England and Wales in 1911

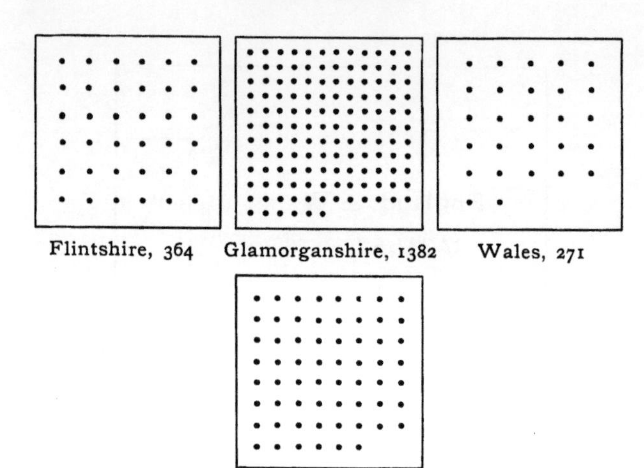

Fig. 3. Comparative Density of Population to the square mile in 1911

(*Each dot represents* 10 *persons*)

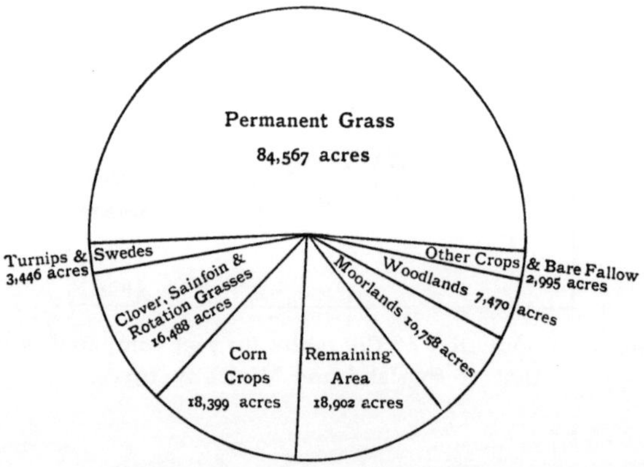

Fig. 4. Proportionate Areas of Land in Flintshire in 1912

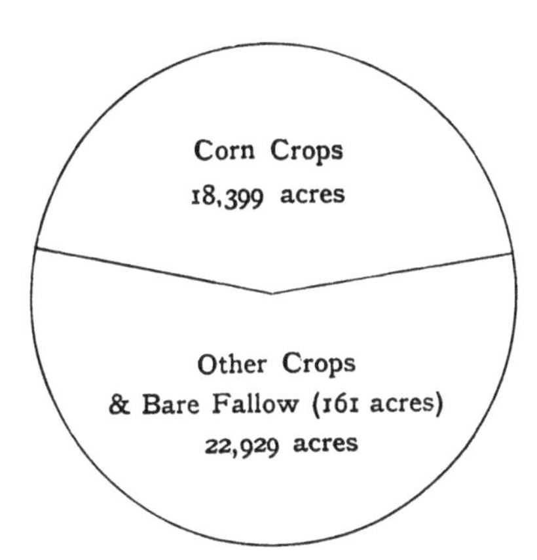

Fig. 5. Proportionate Area under Corn Crops in Flintshire in 1912

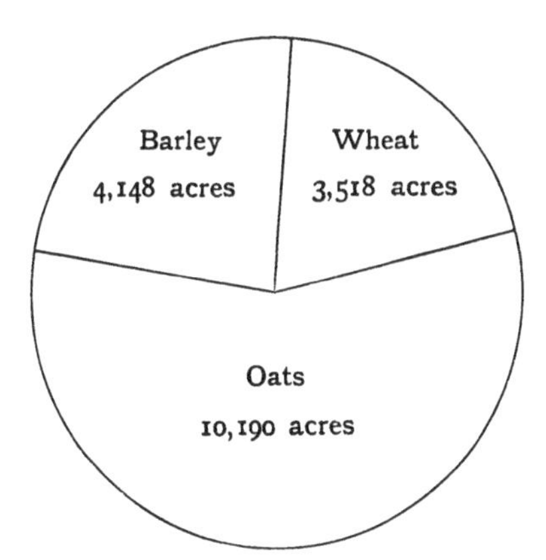

Fig. 6. Proportionate Areas of Chief Cereals in Flintshire in 1912

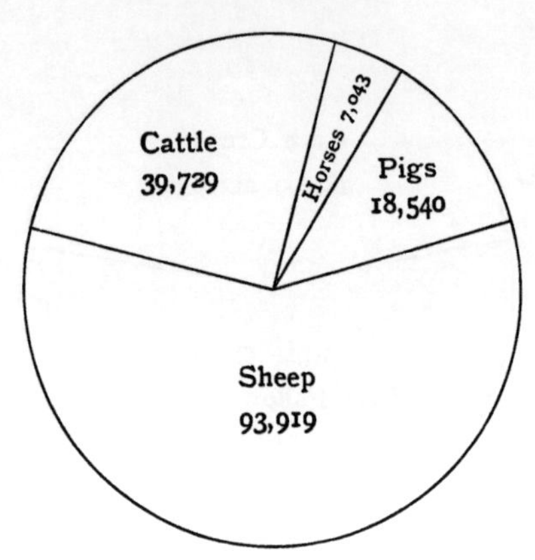

Fig. 7. Proportionate numbers of Live Stock in Flintshire in 1912

www.ingramcontent.com/pod-product-compliance
Ingram Content Group UK Ltd.
Pitfield, Milton Keynes, MK11 3LW, UK
UKHW042143280225
455719UK00001B/73

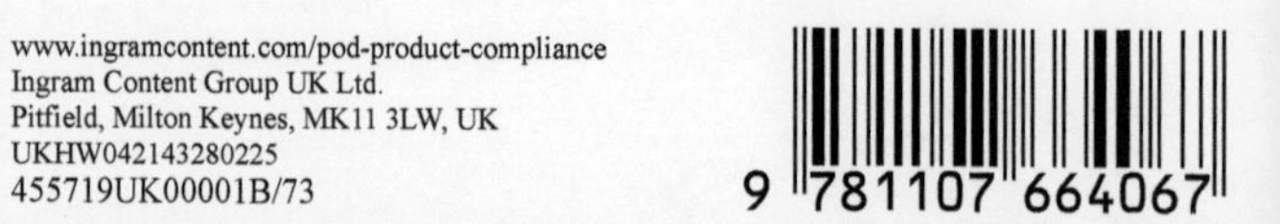

9 781107 664067